Author's Note:

Innovation Engineering Version 1.0 – Beta Release
(Typos included at no extra cost)

You are currently reading the Version 1.0 of this book. In the iterative spirit of this book, this version has been released in advance for the many aspiring innovators who have expressed that they can benefit from this work even in an earlier state. We expect to add additional chapters, cases and updated illustrations in the next release. However, by providing an advance circulation of the book, we expect to offer more timely help to the innovators, entrepreneurs, and executives who would prefer to use it sooner rather than later.

Ikhlaq Sidhu

Publication Team:
Ikhlaq Sidhu, Author
Keith McAleer, Communications
Kyle Giffin, Editing
Elias Castro Hernandez, Print Publishing
Victoria, Wu, Graphics Rendering

For Jyoti

Foreword

Over the past several decades, we have entirely revolutionized our understanding of entrepreneurship and technological innovation. Widespread exponential progress has allowed innovation to expand beyond the start-up world and into the wider sphere of society, affecting the lives of every-day people. These recent developments are advancing our world and changing lives for the better.

Thus, we are brought to the purpose of this book. This book will help you to effectively create and innovate, building upon lessons developed in Silicon Valley and at the University of California, Berkeley. Whether this creation is a new company, a start-up within a larger firm, a government project, or new technology, the principles outlined here provide a practical framework for innovation and execution. Using these principles, you will be enabled to more successfully create the product, service, or technology while effecting positive change.

However, it is no secret that most innovative projects still fail to accomplish their goals. The vast majority fail to even launch in spite of the fact that teams are generally intelligent and highly skilled. So what problems are they running into? What missing piece could give these teams the edge to success? The answer is a proper framework for ideation and execution. This book lays out the roadblocks faced by creators, providing solutions in a manner that is easy to digest and simple to implement. Cutting out the fluff, this book paves a straight forward path to execution.

Ikhlaq Sidhu has been a colleague and partner of mine for over 15 years, working extensively with me to develop innovation programs at UC Berkeley and beyond. I feel very fortunate to have collaborated in support of this work since the very first days of the creation of the Berkeley Sutardja Center for Entrepreneurship & Technology. This includes the days before the founding of the Center, at a time when former Dean Richard Newton and I were discussing who to recruit as the right person to lead this new area of entrepreneurship and technology when it was being conceived at UC Berkeley.

The concepts written in this book have been iterated, tested, and improved throughout a massive journey at UC Berkeley that includes hundreds of technical innovations, new venture creations, and corporate research projects. I am excited to finally see it all come together in the cohesive format of this book. Even more exciting is the knowledge that people around the world will now have access to these principles, allowing themselves to work more efficiently and effectively in an effort create, innovate, and change the world.

Tom Byers
Faculty Director, Stanford Technology Ventures Program (STVP)
Bass University Fellow in Undergraduate Education
Entrepreneurship Professorship endowed chair, School of Engineering
Stanford University

Table of Contents

Acknowledgments

Before we go any further, this book would simply not have been possible without the contributions of so many executives, technical leaders, entrepreneurs, colleagues, and friends. Of course, this includes the innovators in the cases studies Richard Din co-founder of Caviar, Imprint Energy co-Founder Christine Ho, Nir Merry from Applied Materials, Andrew Laffoon co-founder of Mixbook, Jayanta Dey at VMWare, Michael Shebanow IEEE Fellow and engineering leader, and Luke Kowalski Oracle executive and Berkeley instructor. I would also like to add thanks to executives who have taught within my programs including Charlie Giancarlo former chief product officer at Cisco, Mike Olson co-founder of Cloudera, author & entrepreneur Steve Blank for his many supportive conversations, Rich Redelfs formerly Foundation Capital, 3Com, and Qualcomm, Jerry Fiddler founder of WindRiver, Michael Marks former Flextronics CEO, Jim Davidson co-founder Silver Lake, Charles Fan former VMware executive and entrepreneur, Tesla co-founder Marc Tarpenning, venture capitalist Shomit Ghose, Mai Le from Uber and Yahoo, and many more. I can recall at least one important concept from each of them that has been synthesized into this book.

Further, adding to this list, are our major benefactors of the Sutardja Center at Berkeley, In Sik and Isabelle Rhee and Pantas Sutardja and Ting Chuk. And thanks to my colleagues including Phil Kaminsky former department chair and associate dean, who actually advised me to write this book, Ken Singer with whom I developed the Berkeley Method of Entrepreneurship pedagogy, our former deans during the development of the Center, Shankar Sastry and Richard Newton, our current Dean Tsu-Jae King Liu for her support during its current growth, IEOR department heads in this period Lee Schruben, Rhonda Righter,

Ken Goldberg and Max Shen, co-founders of the Center Jon Busrgstone & Stacey Lawson, Tom Byers at Stanford for a long term support and collaboration, Paris de L'Etraz at the IE Business School whose expertise includes mindset & comfort zone and with whom I've collaborated on many real life technology business projects, Alfred Tan at HKBU who inspired a global educational application for this text, Bjorn Hartmann as a collaborator and head at Berkeley's Jacobs Institute of Design Innovation, and the team within the Sutardja Center including Jocelyn Weber who worked with me to develop our executive leadership programs, Alex Fred Ojala who co-developed Data-X, Keith McAleer our communications expert, Victoria Howell who continually provides great feedback, Kyle Giffin who led the editing process, Elias Castro Hernandez who led the publishing process, and Victoria Wu who rendered the illustrations.

So now, and with great appreciation, we can now begin with the Chapter 1.

Chapter 1: Summarizing the Problem and Introducing the Solution

If you only have time to read one chapter of this book, read this one.

This book is intended for anyone who hopes to create and build something new that will get used in real life. This includes (but is not limited to): company projects, mission critical applications, revenue generating services, new ventures, government organizations, non-profits, and start-ups. You may also benefit from reading this book if you work in an applied research lab or are a student hoping to have a genuine impact in real life.

The reality is that most innovation projects fail. There are myriad reasons why this may happen. Many projects seem at first to be progressing, but quickly develop into ineffective, complex, or overly expensive solutions. These projects often fail to generate enough internal funding to even continue. Other times, they deviate from their initial purpose, ultimately missing the mark in terms of serving customers and solving the original problem.

Consider the following two perspectives:

The CXO Perspective[1]

You are a leader or chief executive officer within your organization. Looking at recent developments within your company, are you happy about the return on investment

---

[1] CXO means any of the following C Suite leaders such as CEO CTO, CMO, and others.

generated from newer and more innovative projects? Do you think the time and resources going into these projects have been spent in the best ways? Do these projects succeed often enough that they generate significant revenue?

If the answer to any of these questions is NO, then this book is the key to helping the entrepreneurially-minded people in your organization to adapt, innovate, and enter new markets. They will learn to innovate more effectively, more efficiently, and with significantly greater success.

Innovator Perspective

You are an individual aspiring innovator. Do you believe you have what it takes to navigate every obstacle that may impede you? Based on what you already know, do you feel confident that you'll be able to make your new product, business, or technical project legitimately successful? — If your answer is NO then this book can help you and your team navigate the journey to successful innovation. And even if your answer is YES, you might still want to read a little further.

Pathology

The following are examples of failure scenarios in the context of firms:

1. In any firm, there are teams that pursue new products and services, but for different markets and/or with new technologies. The plan is generally logical. The teams are often smart and experienced. However, few of these projects actually succeed, scale and/or generate

significant new business.

2. A variation of the above are the technology concepts,
    often developed organically in skunk-works project. The
    technical expertise is often high, the technology looks
    promising, and often the team has a powerful
    demonstration, however, the technology most likely
    never gets used in real life.

3. An Innovation Center is one type of organization focused
    on developing the latest new concepts at any firm but
    separated from the main company. The leader is usually
    a very capable person who previously scaled a large
    business. The center is typically staffed with highly
    talented technical experts. However, there is often a
    large chasm between this group and mainstream
    organization.

In each of these cases, the teams are highly-skilled, and yet the
failure is common. And, the returns on those investments are
often questionable.

The Challenges with Innovative Projects

The first challenge of any innovative project is that the path is
unclear. If your team has done it before, it is not an innovative
project. In this case, you know what needs to be done, and you
can just do it. This is execution.

In the case of an innovative project, which actually creates
something new, the team must "execute", but the team must
also "learn" at the same time. It is like driving to a destination,

while learning how to drive a car. In any given project, this means the team is learning about the tools, the users, and the environment while at the same time they are developing the innovation, product, and/or business.

There is a second challenge, which only aggravates the first problem. This challenge is that the destination (or goal) may literally be shifting while you are on the journey. In a project situation, this means that world is changing around you: new technologies are being developed, the user's behaviors are changing, new competitors are entering the market, and the entire landscape is shifting.

Finally, all of this execution and learning has to be done by a team that can work together as a high performance group, with trust, and alignment, and cohesion.

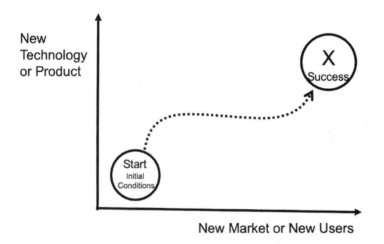

The path of an innovation project must be learned during its execution. Meanwhile, even the destination may shift and yet the team must continue to stay coordinated and aligned.

Myths of Innovation Projects

A common misconception is that these issues can be overcome simply with a combination of good planning and talented people. Consider these example myths which lead to a pattern of ineffectiveness:

(1) **Myth:** A smart, technically-skilled team will be successful.

Not necessarily. ***Execution and leadership*** are crucial components to innovative success.

(2) **Myth:** A logical business plan is the key to success.

Not even close. The truth is that successful innovation is vastly more nuanced than validating a user-need or finding a working business model. ***EQ, innovation behaviors, and mindsets are also essential.***

(3) **Myth:** Lots of resources and energetic activity will lead to new technology or new business.

No: teams often ***confuse activity*** with ***progress***.

Characteristics of Successful Projects:

Having advised and documented hundreds of innovation projects in companies, research labs, and new ventures, we have observed that a team's success is most evidently increased by integrating the following 3 elements:

1. Story Narration

Successful projects start from a "story narrative." This narrative has a wide range of types from contextual business scope to technical prototypes to full-blown venture pitches. This story is the binding ingredient that keeps everyone in the team aligned to work towards the same objective. However, even more important than the story is the way that the story is developed by the innovating team. Many case study examples within this text provide insight and patterns of this element.

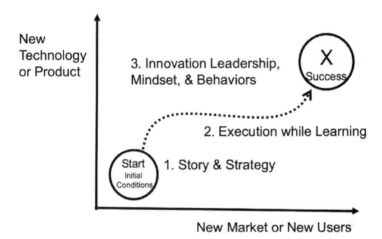

The same path of an innovation project annotated with the characteristics of successful innovation projects.

2. Execution While Learning

Successful projects also require the team to practice "execution while learning." Many teams get stuck on the story or starting point of the project, failing to prioritize the most critical aspect of the project which is execution. These types of teams falsely believe that developing or perfecting a slide presentation is the same as creating the product, business or technology. Despite endless planning and status meetings, the team's pace of active and measurable progress remains slow.

In standard, non-innovation-oriented projects, execution is more straightforward, although it can be somewhat monotonous. In this case of routine or incremental projects, there is a known process for executing, and it is ingrained in the daily activity of the organization.

However, when creating something new, that has not been built before, we cannot simply make a list of steps to do. In these cases, teams must consider many open variables:

- Is the proposed idea technically possible?
- What new skills will the team need to develop?
- What technical platform best fits the team's objective?
- How much will a customer pay for the proposed product or service?
- Will a customer pay at all?
- Does this idea actually even solve a problem?
- How does this fit into the current industry or environment?

In an innovation project, execution is, by necessity, coupled with learning. And in the background, the world is also changing as this project is being developed. Markets, climates, and consumer behaviors are undergoing a never-ending shift. Innovative teams must embrace and thrive under uncertainty. This embrace, however, is easier said than done. Even teams with a track record of strong execution typically struggle to "execute and learn" at the same time. This requires a well-balanced coordination between learning and doing.

The process that we advocate in this text works in synergy with today's approaches to entrepreneurial management, user-oriented design thinking, and agile development.

3. Innovative Leadership, Behaviors, and Mindsets

Technical and business development both require an innovative mindset and a set of behaviors. Leading an innovative team like this is much different from managing a team to execute operational tasks. Typically starting with only a few team members, these innovative teams benefit from adding innovative behaviors such as generating trust, broad thinking, wide comfort zones, inductive learning abilities, and emotional intelligence. Innovative leadership requires a curated selection of qualified people and a constant reinforcement of innovative behaviors as part of the ingrained culture of the team. Leadership further requires some level of influencing, coaching, and project scoping. And further, a leader must continuously strive to develop alignment among all the members of the innovative organization in order to achieve success.

Included throughout this book are case studies to demonstrate the model of innovation in different environments and stages of project maturity. The leaders interviewed have used and emphasized variations of innovative behaviors, mindsets, and leadership techniques, providing real-life examples for the reader to better understand the concepts. Meanwhile, this book provides practical, step-wise guidance for any team and its leadership to develop into an aligned, high performing, and innovative organization.

Innovation Engineering

In this book, we introduce Innovation Engineering as a solution for creating just about anything new. Innovation Engineering is an approach to make innovation projects more effective, efficient, and successful, even in technically complex areas. Innovation Engineering includes the successful characteristics of innovative and entrepreneurial projects listed in the last section.

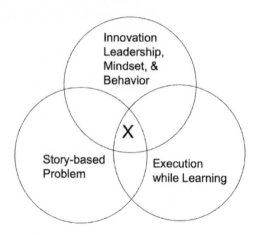

At its core, Innovation Engineering is the result of using the approaches, processes, behaviors, and mindsets of entrepreneurs/innovators within the context of technical and business projects.

Book Organization

In our experience, one of the best ways to understand and communicate innovation within a team or organization is with a pyramid of layers as shown in the figure below.

We are using this layered framework to organize the contents of the text. This framework, can also be used to increase the success rate for innovation projects as well as diagnose success and failure patterns.

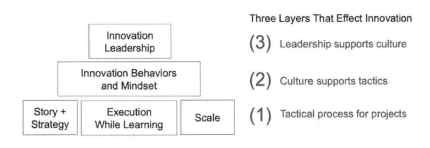

A pyramid framework that allows innovation projects to succeed - and also allows firms to adapt

The lowest layer is the tactical process includes **story/strategy** and **execution while learning**, and ends with **scaling**. The full sequence of the tactical process is covered in Chapters 2, 3, and 4. Additional depth about story and strategy is provided in Chapters 9 and 10.

A second layer is the behaviors and mindsets of all the people in the team or organization. This culture support the tactical process. If the culture is the wrong culture then it will instead slow the innovation process. The matching **behaviors and**

mindsets within the team culture as explained in Chapters 7 and 8.

The top of the pyramid is the leadership style that is needed and tuned for Innovation. Leadership creates the culture in the organization which then supports the innovation at the tactical layer. Our model for **innovation leadership** is explained in Chapter 5.

This material is conveyed using a straight-forward and directly applicable approach. The material is ordered, not from top to bottom in the pyramid, but instead in the order that provides the most value for a project's success.

This text further provides:

1. A practical step-by-step guide to genuinely understand how innovation actually works so that your team can know what to expect and how avoid common mistakes made by less informed innovators.

2. Real-life case study examples, with people from some of the world's most successful innovators, provide context, patterns, and playbooks — retold through our personal interactions.

3. A holistic and validated method that does not replace any existing tools or processes, but instead integrates the most current of innovation practices (such as design thinking, entrepreneurial management, and agile development).

4. Leadership advice necessary for any manager or entrepreneur to build and manage a high-performance innovation team.

5. The behaviors and mindsets that need to be reinforced within any innovation project.

> *Our validation tests indicate that the teams which properly use the approach and tools of Innovation Engineering tend to accomplish their innovative projects at a pace which is approximately 4X faster than they would without this process and with higher quality results. They also onboard new team members faster, they have much fewer unnecessary meetings, and they even have a more positive outlook on the project itself.*

How to use the Book

The best way to use this book is as a guide to any team's innovative project. Examples of applications include:

- New products and services
- Government innovation projects
- Technology projects in any organizations
- Innovation Centers
- Applied research projects
- Student capstone projects

The chapters of the book cover each layer of the framework. Along with interleaved case studies, this book, it provides a simple and direct understanding of how to begin, advance, and scale an innovative project. Each case study also follows the exact same structure, demonstrating how to apply it to any project imaginable.

Use the case studies and the framework as a guide to navigating this journey in a wide range of organizations types and maturity stages. Compare what your team has been doing with the framework to identify areas to improve the existing process and culture. This will allow you to add and integrate elements that you have not yet considered within your own approach to innovation.

Don't Throw Out Old Processes

The framework of Innovation Engineering does not replace the existing processes and tools that are already used by innovative organizations, whether those are agile product development or customer development processes. Instead, the model is intended to provide a path for navigation and learning which will amplify the progress and efficacy of any innovative project.

Innovation Navigator — A Critical Tool for Innovation Project Management

Also included is a tracking tool called the Innovation Navigator. This tool (illustrated below) is based on a spreadsheet template to be shared with all working members of the project. Again, this tool is not intended to replace existing tools. Its main purpose is to reinforce *inductive learning* within the execution of the project. Inductive learning is the ability to learn what to do next from the current environment or based on results from past actions. It forces the team to "think reflectively" on critical topics, preventing them from charging ahead on work that is not helpful.

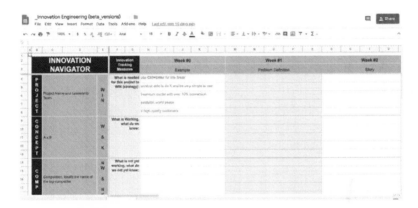

The Innovation Navigator, a spreadsheet based tool to reinforce inductive learning within the execution of the project

Some Relevant Background

This book has been written on behalf of the many technical leaders, students, entrepreneurs, and executives that I have had the fortunate opportunity to interact with over the years. The following experiences and opportunities have laid the groundwork for this book.

Engineering Leadership: Since 2011, my role in creating and teaching in the #1 Engineering Leadership program in Silicon Valley has provided me the opportunity to share experiences with hundreds of executives and technical leaders in some of the world's most innovative and reputable firms. These firms include Google, Apple, Samsung, Cisco, Bosch, Applied Materials, Lam

Research, Yahoo, Network Appliances, Qualcomm, PayPal, VMWare, Broadcom, Bosch, Volvo, GM, Porsche, Honda, Kellogg, and many other cutting edge companies.

New Ventures: Over the past 15 years, I have used my previous experiences in innovation and entrepreneurship to also advise countless startup ventures. This exciting work has led to my capacity as the founding director of the Sutardja Center for Entrepreneurship & Technology (SCET), the only prominent center of its kind at UC Berkeley. Other aspects of this book draw from the Berkeley Method of Entrepreneurship, a globally accepted approach to developing people at a psychological level through studying the behaviors and mindsets needed to create successful new ventures.

Technical Projects: Finally, my experiences are colored by having participated as a team member and advisor in literally hundreds of complex technical projects, state of the art applied research lab projects, and high-stakes business challenges. In 1999, I received the 3Com Innovator of the Year Award at a time while heading a large part of the company's advanced technology labs. During that time, 75 patents were granted to me and my team in areas ranging from internet communications to handheld computing. Even today, in the Data-X course at Berkeley, my students implement state of the art projects using machine learning tools and data science concepts for projects that predict winners of basketball games, separate paper from plastic, diagnose medical conditions, create innovative data pipelines, and much more.

Putting It Together

The result of all these experiences is this book. It has been developed to synthesize the fundamental factors that have contributed most to the success and impact of innovation projects.

Furthermore, this book aims to help the reader integrate all of these concepts into a practical, usable framework that can be applied to any innovative project. It *decodes* innovation, making it accessible to anyone. Further, this text documents provides real-life approaches from innovators, entrepreneurs, executives and technical leaders. With this repository of case studies and key innovation factors, our goal is to help people around the globe change their own parts of the world by making their make early stage technology projects predictable and successful.

Post Chapter Note: Symptoms of a Broken Innovation in the Mainstream Organizations

Innovation process can also be broken not only in the "start-up within a larger firm," but also within the mainstream activities of the firm. How can you tell if the approach in this book will be helpful within mainstream product development? Let's start by looking at common symptoms that indicate the team has some misconceptions about what will make your next innovation project succeed.

1. *Organizational indecision about new products and services.* This may stem from team's inability to decide, which then manifest into disagreements or tension between marketing, product management, and development teams.

2. *Delays and halts in the progress of the project.* Projects often receive extensive internal investment but still end up failing.

3. Lulls in productivity, leading to a slower rate of product and service delivery.

4. *Poor or underdeveloped technology strategies.* This may be the selection of an unsuitable technical platform, the wrong partnerships, or poor suppliers. This can result in unnecessary complexity and ultimately in the inability to compete.

5. *Team morale and retention issues*, causing the best people in the organization get frustrated and seek other opportunities.

These symptoms generally progress to an illness within the organization which results in poor competitive performance, causing the overall business to weaken and suffer.

Introducing the First Case Study

In the section that follows after this page, we will look at the first case example of an innovation project or venture. As you read the short case, as well as future cases in this text, think about these questions:

1. Do the innovators/founders follow any particular process?

2. From where do they start and how do they proceed to narrow down the concept and then execute the project/venture?

3. Is there a particular style to the leadership of the project?

4. Are there behaviors and mindsets allow the team to be successful?

Looking at all the cases from this point of view will help you understand the framework of this book.

Case Study: Richard Din and Caviar

Bootstrapping a New Venture

Entrepreneur Richard Din launched Caviar in 2012. His goal was to provide a food-delivery service for dine-in restaurants that didn't normally deliver. After just two years of operations, Caviar had employees and service in New York City, Seattle, Washington D.C., Los Angeles, and San Francisco. This rapid growth led to Caviar's eventual acquisition by Square (SQ) in August of 2014 for over $100 million USD. Today, the services provided by Caviar are used by more than 50,000 restaurants across the United States.

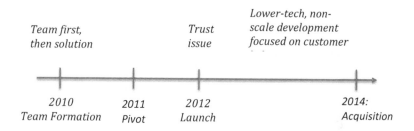

Caviar's startup journey

The Initial Conditions

Richard Din, an electrical engineer by training, had been working at Electronic Arts for 2 years after his graduation from UC Berkeley in 2008. At this time, he started to work with his co-founders Tony Li, Andy Zhang, and Jason Wang, all of whom were still Berkeley students or had recently graduated. Shawn Tsao and Abel Lin were added to the team shortly afterwards. The group found each other based on shared interests and decided to join forces even before they knew the purpose of the company that they would eventually be starting. One of their first decisions was to settle upon the food and restaurant industry as their target. This decision was based entirely on the passion and interests of the team. Then, having developed an initial concept related to a food coupon solution, they each pooled together $5000 to bootstrap the firm.

> *Team first.* The team first came together, then agreed on a target industry, and finally selected the company's purpose.

> *Passion is crucial.* According to Richard Din, "the passion for food within the team was a key to the team's success." Without that passion, the team would never have even had the desire to order from those restaurants and to persevere when there were tough times.

Story

While the initial concept of a daily-deal food coupon system quickly failed in late 2011, a viable business model based on paid food-delivery emerged soon thereafter. This idea was born of a simple problem noticed during the ideation phase: The Caviar team would often meet and collaborate in the office, and during these times they frequently wanted to order lunch from their favorite restaurants in San Francisco. More often than not, they were confronted with the sad reality that these places could not deliver food to them. So, they started to think whether they should create their own delivery service that *could* deliver from these places. As a result, they became their own first customers.

> *The initial story was the articulation of the business model. The team had agreed that delivery would be scope of the business. The rest was a demonstration of the product from the user's viewpoint.*

Another key factor leading to the delivery concept was the following important insight: none of these restaurants had incorporated delivery into their business models. To these businesses, delivery may have been a consideration, but one which required financial backing and an entire logistical process that was simply too distracting and complicated. They were more concerned with producing high-quality food and dining services. In the eyes of Caviar, this demand for (but lack of) delivery was a problem waiting to be solved.

The Caviar team now had two positive drivers to work with: (1) they wanted to use this type of service themselves and (2) they believed that restaurants would be happy to outsource this aspect of the market, as it would also mean more focus for themselves. With some progress made, the concept was now solid enough to be tested in real life.

The first set of success factors that needed validation were as follows:

- Service: Would other customers besides themselves also order delivery from high-quality restaurants?
- Cost: Would these customers actually pay $20 for delivery?
- Viability: Would this service work in the real world with existing restaurants?

Caviar's Use of 'Execution While Learning'

The first challenge was to validate whether or not there was demand for food delivery from high-quality restaurants. In parallel, the technology had to be created to support this validation.

One of the most powerful methods to develop the technology of this venture was to begin with a low-tech approach focused on the user experience. While most teams with this mission would start by developing a comprehensive website to accept food orders, this team did something very different. They instead made the most technically simple interface possible, with only one restaurant and one item that could be ordered. The site's main purpose was to test the behaviors of customers; the team studied how customers navigated the online interface and how their overall experience was.

On the back end, development was even simpler. A person would receive the order, then manually dial the restaurant to confirm order details, drive to pick it up, and deliver it to the correct address. Payment to the restaurant would happen by check later on in the month. A completely manual and utterly non-scalable process. However, the purpose was to know *how* the customer would use the system, not yet to make scalable technological advancements. Only once the user's behavior was known did they start to add the needed technology on the back end of the website.

> *Many projects go wrong because they focus on technology, scale, and high-quality implementation too early on.*
>
> *At Caviar, the original web service was intentionally manual and blatantly unsophisticated. The first objective was to get the user's behavior and business scope correct. Only after that did the technology and scalability factors become important.*

This view is contrary to most technical teams who often fret about scale and technical perfection far too early in the process. In the case of Caviar, the user experience and scope of the business were first studied carefully, and *then* methods to scale and perfect the process were implemented at a later time.

Leadership and Innovation Behaviors

Trust: Among the most key leadership challenges that any team might face is the resolution of whether to remove founders or early employees. This difficult decision was made by the Caviar team after some struggles with some of their earliest members. It began with sloppy and error-ridden work being produced by these team-members, issues which were amplified as trustworthiness began to come into question. These roadblocks led to a waste of both time and resources, losses which the firm could simply not afford to incur during its crucial development period. When trust towards a common objective cannot be restored, removing a founder or early employee can be a necessary action for the success of any firm. Such was the case with Caviar.

> *Trust appears as a key issue in the development of any team from the beginning.* *Caviar started with 4 cofounders and several early employees with a strong basis of trust among all of them. However, as trust issues began to surface with some of the members, it became too large of a factor to ignore. Ultimately, the decision to remove some of the team turned out to be critical to the success of the firm,* **as trust was then restored among the team.**

Eagerness to Learn

According to founder Richard Din, the large appetite and eagerness of the team to learn new things was a critical behavior needed for the firm to be successful. Without this, the rate of progress and level of execution would have been highly off-target.

Leadership starts with hiring

The Caviar team also knew that good leadership was largely a function of good hiring. Through meticulous and careful hiring of well-rounded candidates, Caviar was able to continuously develop the trust, technical skills, and innovative behaviors of every team-member within the organization.

> *Author's note: In this case, we have emphasized the concept of delaying technology and scale investment until the user's behaviors have been validated even during the execution while learning phase. We note that some firms today, like Google, have chosen to use a globally scalable perspective from the beginning of their innovation projects, which may be more of a special case for certain types and sizes of firms.*

Chapter 2: The Solution

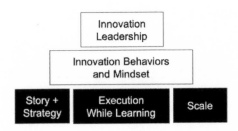

Now let's consider the solution, Innovation Engineering, which we formally define as follows:

Innovation Engineering is defined as a method for solving technology and business problems for organizations who want to innovate, adapt, and/or enter new markets using expertise in emerging technologies, business models, innovation culture, partnerships, and networks.

1. Innovation Engineering is a method that includes a process.
2. The method is used to solve technology innovation and/or business innovation problems within organizations.
3. Often, the goal of the innovation is to change, adapt, and/or enter a new market.
4. These innovations typically involve new or emerging technologies.
5. These innovations may include changes in business model or changes to way the mission of the organization is fulfilled.

6. The method is not conducted in isolation, but instead leverages partnerships and networks of people.
7. The method is complemented by a culture of innovation that includes behaviors, mindsets, and a corresponding leadership style.

At its core, Innovation Engineering is the result of using the approaches, processes, behaviors, and mindsets of entrepreneurs/innovators within the context of technical and business projects.

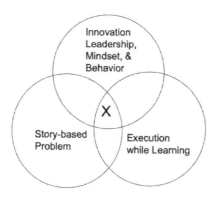

The key attributes of Innovation Engineering in a Venn Diagram Format

The Back Story

In 2005, we launched the Sutardja Center for Entrepreneurship & Technology[2] at UC Berkeley. For the first few years, we primarily used Harvard Case studies to teach entrepreneurship skills to student teams of engineers, scientists, and majors from

---

[2] The center was launched in 2005 and formally named in 2015 by benefactors Pantas Sutardja and Ting Chuk

across our campus. These studies would then be complemented by a classroom experience featuring the *actual* leaders and entrepreneurs who had written the case studies that we focused on. Over time, we discovered something that was key to the evolution of our program: the actual people were more impactful than the case studies. Their behaviors, mindsets, and personalities were fundamental to teaching entrepreneurial capabilities at a psychological level. This led to an approach called Berkeley Method of Entrepreneurship (BMOE) where we would work to integrate the mindsets, behaviors, and psychology of entrepreneurship and innovation into our students — doing so *while* they were in situation of developing a venture within a focused area. For example, the entire class would work in a narrow venture area such as blockchain applications, medical devices for the third world, computer security improvements, or other emerging industry segments. During these technical developments, we worked in the background to add an educational basis that included ***inductive learning, story generation, team formation, and entrepreneurial behaviors into the students***.

In parallel, we realized that implementation of viable technology was just as important as story generation and mindset. Having realized this, we then designed new technical courses focused on technology, application, and implementation in an integrated manner. This contrasts with theory-only topics, in a pure manner without considering aspects of the application. This idea led to the first technical course of this type known as Data-X. Data-X was intended to be the course that anyone would want to take if they wanted to apply these skills in practice.

The really unique thing about this course was that the entrepreneurial processes we were teaching in challenge labs could also, for the first time, be brought into an otherwise completely technical course. The results were staggering. Projects of significant complexity were being completed and demonstrated in 12 weeks, rather than 2 years. The students were ecstatic with their own outcomes, both in education and real-life employment opportunities. Of course, what followed were more technically deep, project-oriented courses of this type. Learn more about Data-X in the Case Study entitled "Data-X: A precursor to Innovation Engineering."

Over the years, two tracks began to develop: One focused on mindset, behavior, team formation, and story development, and the other focused on technical tools, architecture, and implementation.

Our Operational Insight

After reconciling our experiences with students, corporate labs, Silicon Valley executives, technical leaders, and new venture strategists, we came to the following conclusions:

1) Successful projects start from a broadly defined "story narration."

2) Successful projects require "execution while learning."

3) Both the technical and business portions require an innovative mindset and a set of behaviors.

It was this combination of insights has led to the concept of Innovation Engineering.

The Innovation Engineering Process

The graphic below shows the flow of tactical processes within the Innovation Engineering framework. In this chapter, we provide an overview. In the next chapter, we provide a step by step guide so that you can do it yourself. The processes shows 4 phases: (1) the initial set of conditions under which the idea (and team) is formulated, (2) the story or narrative through which both the idea and the team evolve, (3) the phase of execution in an adaptive manner while learning and changing plans, and (4) the phase of scaling the project.

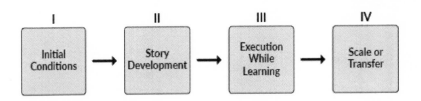

Figure: A High-Level Process Flow for Innovation Engineering
Every successful innovative project can be split into this set of phases. The case examples introduced in this text have also been structured in these same phases.

Phase I: Initial Conditions

No project starts with nothing. For every potentially innovative project, there is always a set of initial conditions. These conditions include the people and background that will lead to the project's initiation. For example, often times a team has already created a demonstration, identified a customer need, or formulated a working business model before the project has even officially begun. In a larger firm, the management may have already identified a vision or strategy for the new product or group.

The most important of these initial conditions is the composition of the starting team. At this early stage, the team is still not fully complete. The team might just be a founder and co-founder, or it could be a whole group working on the project. Either way, if the project is going to be successful, the team must be open to adding new members. These new members will bring a more diverse skillset and background, helping to fill the gaps in knowledge and expertise faced by the initial team. Additionally, it is important to note that the initial conditions not only include the members of the starting team, but also their individual skills, interests, background, networks, and assets.

Equally important is the current environment. This includes the state of the changes generally going on in the industry as well as parallel changes in the behavior of people. As we will discuss, the initial conditions lead to the eventual story that will be developed.

Phase II: Story Development

The second phase of any successful innovation project is the development of a story or narrative. Effective projects always begin with a story or narrative.

It is easy to begin a project by creating a list of requirements, and many team will believe this to be the most efficient approach. However, when a project is not founded upon a story or narrative, it tends to go off-target fairly quickly. This is because the project definition is likely to be already flawed from the beginning, either by being too broadly defined (*i.e.* too vague to develop) or being too narrowly defined (*i.e.* not aligned with the intended purpose of the project).

The issue with jumping directly into development is that the team has not yet established consensus about the target market space, nor have they validated the proposed problem, approach, or solution. Remember that the story narrative not only communicates the problem and solution, but it is also intended to collect stakeholders, generate resources, and prove initial validation for the project. In our experience, there is no better way to achieve alignment and attract resources than by testing a story.

It is still important to understand that there are many formats of stories. In some situations like the Caviar case study, the story was simply the agreement of the scope of their business followed by a simplified, demonstrable product.

They had agreed that they would target restaurant delivery. They felt that customer restaurants would not feel threatened or even care if another firm offered delivery. This was unlike the Mixbook case study where the team could not make their first project succeed because the yearbook makers were indeed threatened and would not support their concept.

Caution 1: It is easy to want to skip the story portion and just start writing requirements. However, the extra time invested in developing and aligning the story will prove very valuable, saving 10x the effort during the execution phase. Without proper story alignment, extra time gets wasted on unneeded features, problems that don't need to be solved, and delays due to general misconceptions about the purpose of the entire project.

Caution 2: We will bring this topic up again, but the story does not need to be perfect when the project starts. The story itself can evolve, however, certain strategic mistakes in the story can really slow progress or even kill the project. We will discuss this later in the chapters devoted to Story Narration, Strategy, and Common Errors.

Phase III: "Execution while Learning"

The next phase is Execution. In this phase the product, service, organization and technology are developed together to a point that they can be easily scaled.

In real life, **Execution** is by far the biggest factor in regards to success, yet it is often ignored in the classroom. High performance execution is not commonly achieved in academic innovation projects or the early stages of business projects, however it is the most fundamental activity for real growth. When evaluating a team, examining their execution is paramount; two teams with identical ideas may have radically different levels of true progress. Some might claim that good execution is simply a result of hard work, but this is a fallacy; there is much more to it.

Execution in this case does not mean that we make a list of steps and execute they one-by-one. This only works when we are able to identify all the steps in advance that will result in guaranteed success. In the case of a venture or early stage technology project, we must execute *while* learning and experimenting.

The reason we must "execute while learning" is that many variables are still unknown; the correct path not yet paved, therefore it must be discovered during the advancement of the project. Often times, the destination of this path changes significantly as the project develops.

Parallel paths begin to form in the Execution phase following the initial story development. In this "execution while learning" phase, we will see how these paths begin to parallel one another:

- Developing the project, venture and/or organization
- Develop the product, technology, or capability using an Agile[3] process
- Evolving the strategy for success in a constantly changing background environment

In the next section we will break down the elements of the process in detail.

Phase IV: Ready to Scale

This is the final step within the scope of this text. All of the combined work in the Innovation Engineering process should lead to point where the project is ready to scale. Generally, this means that there is now a working business model for the product or service, particularly if the project is intended to be revenue-generating. Alternatively, if the project is within a government organization or falls deeper inside a large institution, then the goal is not to establish a working business model, but instead to demonstrate that the mission of the project and/or the mission of the organization has been achieved with small to moderate success. Once this goal has been met, it is time to move on to scaling.

Now that the project is ready to be scaled, we must find the best approach to do so. And in order to scale, it is often true that a

---

[3] For readers unfamiliar with Agile, it is development approach in which requirements and solutions evolve through the collaborative effort of self-organizing and cross-functional teams and their customers/users.

transition to a new team is needed. Typically, scaling is not a process that is learned while executing. The best way to scale is to do it with people who have done the same type of work before. Therefore, scaling is most effective with a team or group of individuals who have expertise & experience.

In this final phase, once a working model has been achieved, holistic problem/solution has been validated, and the solution is also robust, then, and *only* then can the focus be switched to scaling.

Chapter 3: A Step-by-Step Guide to Innovation Projects

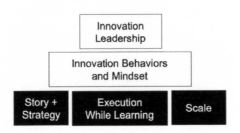

This section serves as a step-by-step guide to starting and successfully completing an innovation project. In the last chapter, we outlined 4 major phases of an early-stage innovation process. In this section, we break those four phases down into eight steps which can be done in sequence. Let this be a practical list of steps that make up a process for innovation. In future chapters, we will add more context including culture, mindset, behaviors, best practices, case examples and other resources. For now, to get started, we will introduce a skeletal process upon which other concepts can be layered.

You can review these steps first with the intention of understanding the actual way that an innovative project starts, grows, and succeeds. You are invited to review the sequence and compare it to the processes that they already use. And after that decide whether it is better to adopt the entire sequence of steps or simply integrate ideas into the processes and approaches that you already use.

Step 1: Articulate the Initial Conditions

All projects start with something and/or someone. No matter whether it's corporate innovation, clinical research, or a classroom project, every project has initial conditions. One might think that a project starts with an idea or requirements, but in reality, it starts with people or at least one person. While projects are often started by a single person, they could also begin with several friends or colleagues who have an interest to fix, change, or impact some aspect of the world. The desire of the people comes first, and then they must start to take inventory of the skills and resources that they already have.

The very first step is to articulate the starting point with the following questions:

1. Who is the initial team?
2. What insight does the team have about a market, technology, or problem?
3. What if anything has already been accomplished technically (a demo, feasibility, etc.)?
4. What if anything has been accomplished contextually (validated problem/solution, initial concept, committed partners or customers, etc.)?
5. What background does the initial team have that makes them the right people to work on this innovation?

What has recently happened in the context of customer behavior, technical advancement, or regulatory changes that make this the right time to work on this project?

Beginning with the first question, let's consider these initial team examples:

a. One or two students decide they want to start a company.
b. Two co-workers start to discuss a gap in the market of their firm.
c. A small number of government employees think that they can better serve the organization's mission by implementing a new technical solution that uses data, AI, and robotics.
d. Two researchers have come across a result which they believe could have value in a commercial venture
e. An executive and staff member starts to discuss a major disruptive threat to their business and begins to brainstorm approaches to defending against this threat, possibly utilizing a new technology.
f. An executive sponsor wants to create a project in a new direction and is starting to build a sub-team that works under him/her to further develop this initiative.

All of these are simple examples which are intended to provide a visual for what the original team may look like.

The initial conditions will start with the original team, but also include other factors about the initial team's background, resources, skills, and networks. First things first, this initial team may have already done some work in the past which leads to the formation of their idea such as a past project giving them insight into a new problem. As discussed previously, even with projects that have not yet moved forward, there may already be a demonstration, proof of concept, proposed business model or other work that has already been completed.

Next, the initial person or team should have skills or intrinsic interests that allow for them to work effectively in a specific area. Skill may range from deep technical knowledge or sales and marketing capabilities. An often-overlooked skill is that of social engagement capability, or emotional intelligence. One's ability to understand and deeply connect with people has profound impacts on their potential for success. Finally, the original person or team should either have, or begin to develop, networks of people that they know — customers, managers, investors, and partners. Curating this network will be key to finding success in your venture.

Step 2: Scope the Concept

The next objective is to narrow down the scope of the innovation concept. The team will, of course, brainstorm many ideas over time. Successful teams might do this process over weeks, months, or some cases even years in the case of researchers who persistently work toward a world changing innovation concept.

One of the most common starting points is that of a "High Concept" which is intended to articulate the initial innovation concept as a combination of two existing and understood concepts. In the figure below, within the Venn diagram 'A' could be a firm like Amazon and 'B' a new market or geography like China. Therefore, the intersection of these two (A x B) results in Alibaba. 'A' is almost anything that has already been proven to work as a business or solution in the past. 'B' is a change in market, new technological approach, or some other factor which makes this concept new.

As an example, Innovation Engineering itself is the combination of concepts that existed before. In this case, let A = IDEO, a successful design firm which developed a process and culture for user-oriented design. Let B = the processes related to innovation by technology entrepreneurs focused in early stage innovation. The intersection of A x B eventually developed into Innovation Engineering, after adding 30 years of our early stage innovation experiences and observations into the mix. Again, there are other ways to scope the concept and/or start the initial story, but this High Concept is one common approach.

In addition to a high concept, the frame of the project should also include an understanding of the competition. In other words, "what will be replaced" if the project is successful. It might be a **direct replacement**, which is a better product of the same type, (*e.g.* iPhone as a replacement for Nokia), or it can be an **indirect replacement,** which is an alternative to the competing product, (*e.g.* LaCroix as a substitute for soft drinks). Either way, it is critical to know what or whom is being replaced. This is highly beneficial, as it provides a focusing factor which helps to prioritize work. It is common that innovators often choose their competition even before they know their competition. In the case of Netflix, founder Reed Hastings felt that Blockbuster Video had poor customer service and was somewhat of an outdated, slow dinosaur. Therefore, he chose to target them and their market long before he knew which product or service would actually be most competitive. By choosing a target, the team can rally against the competitor in order to gain internal momentum.

It is absolutely critical to a project's defense to articulate the primary form of competition. This is because the people connected with the competing solution will at some point try to protect themselves if your innovation becomes a real threat. Your team should not be blindsided by the competitor, who might retaliate with product changes or with political or legal warfare.

Finally, the initial concept scope must match the team's background. There should be some past project, skill capability, or long term interest to work in the area of the proposed innovation concept. If the innovation concept deviates too far from the skills, background, and interests of the team, most managers or investors will not allocate resources towards it. They may really like the project, but will be concerned that the team is incapable of succeeded, thus leading them to find a team better suited for the project. It is smarter and more strategically sound to re-scope the project to start with something that *closer* to the team, and then over time incrementally grow the scope of the project.

From our decades of experience, we've found that technology projects and ventures cannot progress without a clear articulation of these 3 topics as we have discussed until now and listed below for clarity:

1) *High Level Concept*, *e.g.* A = Banking (a business), B=available only online (a new trend or technology), in this case we are making online banking possible online.
2) *The target of the competition.* This means that the team is aware of the firm or technologies they intend to replace.
3) A match of background of the team and the objective of the project.

Step 3: Further Develop the Story

For any successful early stage technology project or venture, the team develops in conjunction with the story and customer validation. In the last step, we highlighted the importance of an adaptable team and initial story. The next step is to further develop the story.

What is a Story

A story or narrative is a set of words and slides that explain the problem, solution, and important contextual information, as well as technical details about the proposed solution. In this approach, the story has two subcomponents:

A) Contextual story

An Investor Pitch: The contextual story is broad in nature. For a new product or service, it may have similarities with a new venture pitch and the complementary 2-minute spoken version commonly referred to as the elevator pitch.

NABC: Another appropriate framework for a contextual story is the NABC model (Need, Approach, Benefit, Competition) developed at SRI and famously used in the development of Siri, acquired later by Apple. If the project is for innovation within a government organization or public resource, the story can still be similar with some minor changes. In public organizations, there must be some change in focus. Specifically, the business model considerations must typically be replaced with a focus on serving the mission of the organization. Regardless, the contextual story is still important in both private and public settings.

NABC Contextual Story Example
- What is the Need?
- What is the Approach?
- What is the Benefit?
- What is the Competition?

Business Space and Scope: In many sample cases, the contextual story is as simple an articulation of the scope of the project and the logic why it will fit with other players in the industry. This type of story would be a simplified business model.

More stories: Many more story types including the product demo and the Amazon 6-page business plan are explained in

Chapter 9 of this text. At UC Berkeley's Jacob's Institute of Design, DES INV 15 Design Methodology or Sara Beckman's Problem Finding Problem Solving class have been used to engage activities that ultimately yield a contextual story that is grounded in real-world observations and needs. According to Jacob's Institute Director, Björn Hartmann, a key insight from studies of what designers do is that it is often important to develop multiple alternatives in order to understand what the right way forward is. This could involve selecting the most promising of several alternatives or synthesizing a new proposal from bits and pieces of individual proposals.

Rough shape to be chiseled down to fit

Story Development: A simple illustration to convey the concept. We first, find the open space, then we repetitively adjust one of pieces that we already have to make it fit.

How the Story is Developed

When considering the contextual story, one of the most common approaches is to start from the market space and then move to scope. Through iteration, the scope keeps getting adjusted until it is a perfect fit. There are two common models that   work   well:

Model 1: Team to Venture Story, *e.g.* Caviar Case
a) Start here: where does the team have passion? And what do they know?
b) Based on this, choose the industry or space where the team actually wants to innovate.
c) What part of the space is open — or where is there a big challenge — where the team can work on a story narrative? (Most Entrepreneurs think like this.)
d) Iteratively chisel away at the existing story until it fits. This means that it will work nicely within the rest of the market. There will be value, it will be a match with your skills, your customers will not see you as a threat, and finally you will actually be competitive.

Model 2: Redefining the Industry Story, *e.g.* Oracle Case
a) Based on the team or existing firm, choose the industry or space where the team actually wants to innovate.
b) Ignore the current state, and think about what would be ideal for everyone involved. How would it be in a perfect world? The solution may involve partnerships and/or changing the entire industry structure. (Apple is also famous for this type of thinking.)
c) What bold changes could be made to redefine the industry space and solve a larger problem?
d) Iteratively chisel away at the existing story until it fits. This means that it will play nicely within the rest of the market. There will be value, it will be a match with your core capabilities, and finally you will actually be competitive.

This contextual narrative also communicates a relationship between the problem/solution, the team's background, and any recent changes in the environment. In this manner, the story may answer the question of why this team is uniquely equipped to develop this project. Additionally, by connecting the problem to current industry or user-behavior changes, the story can also indicate why this is the right time for this project to be launched.

B) Technical Story

Simplified Product Demo: The technical story explains how the product will look and function. One of the most common technical stories is actually a simplified product demonstration when it is possible to create that in a short time. In this case, the user's view of the product is the most important to preview in the demonstration. All of the features do not need to work. [4]

Paper Demo: In other cases, short of building the demo with code or tools, a technical story may be the same demo but done on paper (a paper demo) or a slide version. Both of these will again highlight the user's perspective.

Low Tech Slide Demo: In our programs at Berkeley, we use a template of approximately 5-10 slides for the technical story, which we refer to as the "low-tech demo." In this case, it is useful to think of these slides as a creation of the product or service in slide format.

- What is the product or service?
- What is the user's perspective? How does the user interact with it?
- What are key components and risk levels?

---

[4] Designers call this a prototype and a lot of the skillsets taught in design courses are around producing the right kind of prototype for a given question or given goal quickly.

- What is the architecture?
- What are the initial milestones and swim-lanes?

For this type of low tech demo in slide format as outlined above, we begin with one or two slides to scope or overview of "what" is being developed. Next is a slide section illustrates the user's viewpoint, particularly including what the user will see and when they will use the product or service (may include screenshots, user processes, and other visual prototypes). Moving deeper into the technical details, the low-tech demo (technical story) shows the key technical components. In our version, we list and color-code the key components with red text for the things that we don't know how to develop and green text for those components that are easy for the team to develop. The technical story should include a high-level technical architecture showing how components connect in order to make a system. The last section is titled "Initial milestones and Swim-lanes." The **initial milestones** are the first few steps of the project, small-scale accomplishments providing a foundation for how to proceed. **Swim-lanes** are an assignment of broad areas of focus to individual team members (such as assigning Person 1 assigned to User Interface topics, Person 2 to Algorithmic Topics, etc.).

It is important to understand that these stories are just starting points. We fully expect them to adapt and change as we progress in the project. The combined story (contextual and technical) is simply the beginning of the next phase, which is a path of both execution and learning converged into one.

Step 4: Plan for Partners and Stakeholders within the Story

Most generated stories place far too much emphasis on building the technical aspects and not nearly enough emphasis on the social aspects — partnerships, suppliers, and collaborations are crucial. For any given idea or concept, the goal should be to build as little as possible. There is a large danger in trying to do activities and/or build things in areas where your team has very little core competence. To waste time on these activities is unnecessary, as you may be able to purchase them or partner with someone to include them in your offering and thus saving both time and physical resources and increasing the likelihood of success. **Don't make the critical mistake of wasting your effort.**

To build everything is futile. To stay with your core competence is smart. Build your story in a way that minimizes the development necessary for launch, and do this by maximizing your collaborations.

First consider the concept you have in mind. It does not matter whether it is a venture, corporate initiative, or government program. First ask yourself: what is the value that your team can truly bring to the concept? Perhaps you have a valuable data-set or you have a unique core skill which is hard for others to replicate. Maybe you have access to a sales channel or highly effective branding techniques.

Whatever your competitive edge may be, it is requisite that you have some core capability which is both (A) scalable — meaning you can continue to provide it as you grow — and (B) hard for others to copy or replicate. These factors are indispensable.

For everything else that is not a core capability, you should seek potential partnerships and suppliers, including those directly in your story narrative. If supply can be bought off the shelf, then mention that in the story. If there is potential for partners to supply key elements, then mention that in your story. It is possible that the key part of your implementation process might only be the glue code that connects other components or services together. The primary requirement is that the full project is **difficult to replicate by others**.

This does not mean that you cannot offer a full solution. The story can still be created so that customer or users will perceive the solution to be completely offered by you, even when partners and suppliers are key elements of the solution.

If a project story is developed with this mindset and the rest of the internal work is completed, the project can be quickly launched and scaled. This will allow for growth within weeks and not years.

Step 5: Getting Started Requires a Context Leader and a Technical Leader

To start the process of execution, there must be one person dedicated as **Context Leader** of the venture or project. A second person is typically required for the agile development, or possibly just the prioritization and/or management of that development. The **Technical Leader** may be the Venture CTO, a hacker, or possibly the Scrum Master (facilitator for an agile development team) in a project within a large technical organization. Often, we refer to the Context or Technical Leader interchangeably with the Context Lead or Technical Lead in this text.

In different organizations, titles may vary. And to some degree, the titles are largely irrelevant. However, the functionality of these two are consistent no matter where the work is going on. The table below illustrates this point:

Organization	Context Lead	Technical Lead
New Venture	CEO	CTO
Large Firm	Product Manager	Technical Manager
Student Project	Project CEO	Project CTO
General Project	Business lead	Hacker, Tech. Lead
Government	Agency Head	Technical Lead

As a shorthand, let us refer to these leads as Context Leader (a type of project CEO) and Technical Leaders (a type of Project CTO). These two people must be able to trust each other and work together with a high degree of communication. There may be exceptions when one person can play both roles in the beginning, however separate roles are highly preferred.

Before we proceed, note that a log or journal must be kept during the execution of the project. We have developed the Innovation Navigator to be one version of this log for managing the project. Both the Contextual Lead and Technical Lead will contribute to the log. The log serves as a record of the actions and considerations in the Learning Journey and Agile Implementation Phase. This is important for the team to be able to reflect, learn, and take the most effective actions during the process.

Step 6. Defining Success

This is another thing that is easy to skip, but important to do. The answer may simply be 3 bullet points defining a vision and measure of success. Some teams have even written a future press releases describing their own success in the future. Examples may include a) sell 1M units, b) become the thought leader in x category, or c) even solve world hunger ☺.

To be more strategic, try to answer the question, **how will the project win** – or – **what would give the project/company an unfair advantage** over competing solutions. The goal must be measurable, and then it needs to be articulated in a few bullet points. It is part of the process that this goal may change in future, but to get started, it is important to define, communicate, and archive your team's understanding of success.

Step 7: Execution

The goal in this section is to make good "Execution" both understandable and predictable. Again, execution, as we define it here, is not a step-by-step process. We must use our previously defined technique of **Execution While Learning**.

As shown, there are two types of Execution While Learning that follow the initial story development. One type is for contextual progress where we develop the entire venture and/or organization. The other is during the **Technical Progress** where we develop the product/service/technology using agile and design approaches. These two are interconnected with one another.

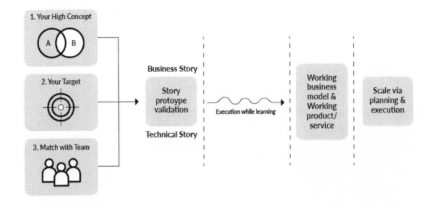

Learning Journey

When it comes to the creating and executing your business or organization, the key function is to learn and navigate while you Execute. This involves 3 components:

1) Separating knowns from unknowns in order to fill knowledge gaps
2) Making continuous strategy adjustments
3) Gathering the people and resources needed to succeed

The three components form a cyclical, iterative learning process. The Learning Journey is typically run and owned by the Context Leader (or product manager in a larger organization). The Technical Leader also continuously cycles through each of the subcomponents below, adding new information, resources, and then finally action items. These iterations can be daily, weekly, or monthly depending on the stage and type of project/venture.

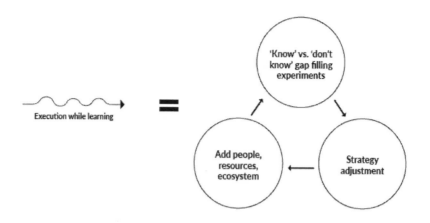

1. Separating Knowns from Unknowns

This component is about listing what you know how to accomplish versus what you are yet unsure how to achieve. These are of course just examples, but the "Knowns vs Unknown" list might look like something this:

Known (or already working):
- i) validated the customer pain
- ii) able to demonstrate a solution
- iii) solution is more efficient than competitors'

Unknown (or not yet working):
- i) validated that the product actually solves the customer's pain
- ii) know that the solution can be scaled for more customers

This process highlights the boundary between knowns and unknowns, which is where we can find the highest impact point for learning and execution. The action required in this process is to *prioritize and select* the next objectives to learn, and then make progress by running experiments. The learning and execution processes are achieved by testing various hypothesis as part of the process.

Based on the results of this iterative phase, various action items for context execution and technical execution will be developed. To make this model effective, it is important to understand that *the action items are **not to be decided first** as in the normal case of execution, they are to be **driven from the separation of knowns and unknowns.***

2. Strategy Adjustment

A second component in the learning process is Strategy Adjustment. This means that the team must regularly ask the question, "How will we win?" The answer may change over time as new information is gathered.

Strategic change is about choosing a variation in the path or destination. If we were actually using a map to plan a trip, the strategic question would be whether we should choose a different route or even try to reach a different destination, based on new information that we discover along the way — such as the changes in weather or traffic congestion.

In a project or venture context, here are some examples of strategic re-evaluation questions:

- What does it mean to win? Maybe our definition of success of the mission has changed.
- Has our intended destination changed?

- Do we want to build a platform versus a product or service?
- Is our current business model still suitable, given what we've learned?
- Do we need to prioritize different features? Where do we derive our new value from?
- Do we need a different mix of team skills? Should we recruit different types of team members, advisors, partners, or customers?
- Has the environment changed? What have we learned from the news, social behaviors, or technical drivers, and how does this change our approach?
- Has new technology been introduced in the market that can be used off-the-shelf instead of building it from scratch?

3. People, Resources, and Ecosystem

The final component of the Learning Journey is to prioritize and collect new stakeholders. Again, this may include team members, advisors, investors, customers, partners, etc. *Note that adding new people can be considered the same as collecting resources or funding.* A new team member, if chosen well, may bring a working prototype along with them. Another might bring channel relationships that are essential for growth. Having the right advisor might give the competitive edge to secure a large funding source or bring on a massive new client. In each case, we should be strategic in the selection of people. Having too many people is can be detrimental, whereas having fewer people with the right assets and skills can be very effective. When all the stakeholders are recruited, the result is an ecosystem that will be needed to grow the project or firm.

To reiterate, the goal of this component is to **use new information** that is learned **to select and recruit new people** to the project/venture and its ecosystem. This same approach should be considered when growing the project/firm and collecting funding and other resources.

Navigating the Learning Journey

Picking from all these different components creates an iterative, inductive learning model that produces results in any new project or venture. Each component brings new information and resources, which are then used in the next step to further reflect and act. These actions may include experiments to prove a hypothesis, the development of a new strategy, or the recruitment of new stakeholders. Along the way, plenty of execution and learning occurs.

In the next section, the text will lead the reader through the use of a spreadsheet-based tool we have used successfully to guide the innovation process. We refer to it as an Innovation Navigator and a screenshot of the tool is illustrated immediately below.

Innovation Engineering (IE) Project Tracking	Define the measure of success for the project in advanc
Use this log to track your weekly progression through the IE framework.	use Ctrl+Enter for linebreak
Project Name:	Success is measured by a working prototype, with validated UI. We will l
Project CEO:	direction.
Project CTO:	

	Example	Week #1
What is needed for this project to WIN (strategy) by CEO:	a) 3 high quality customers b) product able to do X and be very simple to use c) freemium model with over 10% conversion d) establish world peace	
What is Working, what do we know:	* technology works * demo is inspiring	
What is not yet working, what do we not yet know:	* sales pitch not yet effective * will not scale beyond 40 users/day * user interface not simple enough	
Agile Development list for next week (top 3-5) by CTO:	* Fix UI * find different way to scale users	
People recruiting update by CEO:	We just added a new advisor to help build our channel. We need to recruit a volunteer for UI design.	
CEO Reaction (1-5):	4	
CEO Log: Reflection over last week:	I am not convinced that the users are willing to pay for this because something similar is	

Example of Project Tracking in an Execution Phase with an Innovation Navigator

Agile Implementation for Technical Delivery:

In parallel to the contextual learning, an Agile Implementation is the method that is used for technical learning as well as development of the product, service, and/or technology.

Agile is a topic that has been richly covered many times, and it is used in widely in Silicon Valley and technology centers around the world. The objective here is not to re-explain Agile nor its benefits, but instead to point out that the Agile Development process will continually be informed by the Learning Journey described above. In this manner, it is integrated with the Learning Journey as part of the execution process.

To begin the Agile process, the Technical Leader must lead the activities and drive the process. As the team grows, a Scrum Manager might take on the prioritization aspects of this role. This may also happen in coordination from the product management lead when the organization grows larger. For a new venture, it is typically driven by one of the founders who take the lead technical role.

In the appendix of this book, there is a case study on "Starting an Agile Implementation for Technical Delivery" based on our experiences at UC Berkeley. It can be helpful to review as an example of how to kick off an agile implementation based on a contextual story and technical story.

Step 8: Scale or Handoff

The illustration below is one way to visualize the execution path from the team's starting point to a working solution. Progressive examples of technical capability range from research, feasibility, development, demonstration, subsystem testing, system testing, and launch.

On the contextual side, progressive stages of investment readiness include initial insight and creation of story, validation of problem & solution, validation of business model or mission, development of ecosystem, and identification of operational measurements.

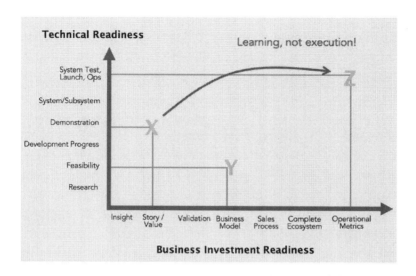

The first question is to know when you have reached the point during which scaling is even possible. In a business or new venture, this point is reached when the business model of the project has been shown to work successfully. The customers are known, engagement is high, sales model is understood, the product or service solves the user's needs, and all the collaborations are in place. At this stage, none of the business model variables need to be adjusted. In fact, operational measures are usually created at this stage to ensure that the business model does not float adrift. As this stage, investing more resources should directly result in larger returns.

If the project is within a public or government organization, then the model of the business should be replaced with the model of the mission. A business model defines how a business earns and retains money. A mission model is a similar model in that it defines the variables, scope, and functions that allow an organization to serve its mission and get credit for that mission.

Note that some cases studies in this book (such as the VMware example) are more focused on the translation from innovation to scale and the issues inherent in scaling.

When a business model or mission model has been achieved, leadership must decide about whether they want to run the project in the next phase. To do so requires acting more like an executive and less like an innovator. Certain teams prefer to hand the project to another industry leader who wants to run the operations of the project, allowing the initial team to go back and start working on a new innovative project. In other cases, the original team wants to grow into roles of operational leaders, and in this case, they must agree to lead the team with less experimentation, fewer changes, and greater stability. This is a personal decision for any team member or innovation leader.

Summary

Presented here is an explanation clarifying how early stage technology and business or mission-driven projects are **launched and executed**. The process presented here begins with initial conditions and ends with the scaling of the project. Recall that our objective is to reduce the high variability in success rates with early stage innovation projects. These processes can be very effective in both academic, government, and in industry settings.

In the next section, we will explain the framework that we have used to offer this step by step process. After the next chapter, we will discuss the culture, mindset and behaviors that complement this process.

Case Study: Oracle Changes Enterprise Procurement

Top-Down Led Innovation that Changed Supply Chains

Oracle Corporation is a multinational corporation centered around database technology headquartered in Redwood Shores, California. Larry Ellison co-founded Oracle Corporation in 1977 with Bob Miner and Ed Oates originally under the name Software Development Laboratories (SDL). By 2017, the firm had revenues of $37B USD and 138,000 employees. The firm offers products based on its relational database technology in areas such as enterprise resource planning (ERP) software, customer relationship management (CRM) software, and supply chain management (SCM) software. Oracle is embedding innovative technologies in every aspect of their cloud, enabling companies to reimagine their businesses, processes, and experiences.

This sample case covers Oracle's entry into the Business to Business E-Commerce, possibly with the first cloud-based Software as a Service (SaaS) application ever developed and used at scale. The case covers the time period starting in 1999, though the business is still dominant today.

> *This case dispels the myth that innovation projects in companies are always organically created from the bottom of the firm. In many cases, the start-up within the larger firm is created by the top executive who provides the original idea, political cover, and guidance through the process.*

This sample case of Oracle highlights an innovation that was driven top-down by Larry Ellison as CEO. The case is told from the perspective of a senior interaction designer in the original project within Larry's team

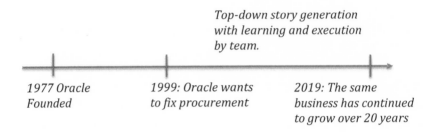

Top-down story generation with learning and execution by team.

1977 Oracle Founded

1999: Oracle wants to fix procurement

2019: The same business has continued to grow over 20 years

Initial Conditions

The problem in 1999 was that purchasing systems in most companies were outdated. The process would typically involve forms like purchase requisitions, purchase orders, and other paperwork that would get faxed back and forth between companies when one firm needed to buy from another firm.

Larry Ellison, who started Oracle twenty-two years earlier, had an initial concept or idea that purchasing would be better if all the firms and suppliers were part of a community that could more easily trade or buy from each other.

This could be accomplished with a website where any firm could log in and buy from other firms that also had accounts in that system. And since Oracle was a database technology firm, it could host that web site as well as all the data about the accounts and items offered by each firm.

> *Because Larry himself had developed the concept and insight around developing a trading community, it automatically provided political cover for the project. And very quickly, this led to a start-up being formed within the existing Oracle which was quickly becoming a larger firm.*

In a departure from traditional business thinking, Larry thought that the suppliers did not actually need to have any of Oracle's software in order to interact with the trading community. Oracle would host on everyone's behalf. And for those customers who did have the Oracle's software, a large marketplace of suppliers would already be available to purchase from.

Based on the contextual story and concepts communicated by Larry within the firm, a product development team was formed. The team would make regular updates to Larry directly. This team quickly grew to about 100 people, with 4 functional leaders (product development, marketing, etc.) and approximately 15 executive stakeholders.

Story

The initial story narrative was the high-level scoping of a business model for how Oracle would play within a new paradigm where a trading community would make purchasing and supply chain management easier and better using the Internet.

Once the team started to form, the story quickly evolved into a demonstration of the product concept. As with other innovation cases, the focus was kept on the needs of the users, but implemented in real life product demonstration.

The demonstrable, minimum viable product was quickly tested both internally within the firm and externally with selected customers like GE. Over time, this product line became known as the Oracle Exchange and Self Service Purchasing within Oracle's E-Business Suite.

Execution While Learning

New technical issues soon appeared as soon as the product started to be tested in live situations. For example, a stateful web server application needed to be developed. Clients would hit a "back button" on the browser causing the system to quickly become confused about the current state of the purchase process.

It was also learned during the execution that it is very helpful to show a status update on every screen for where the user was in the purchase process. This became a "breadcrumb" feature so that people could follow the path back to the place in the process that they needed to execute.

Another key learning that occurred on the path was that customers directed the team to look at the consumer model for Internet shopping. The concept of the shopping cart and checking out was becoming the standard at places like Amazon and other consumer-oriented Internet retailers. The Oracle team borrowed those same ideas for their enterprise product. For the first time, people within the firm could add items to a shopping cart and check out. The other aspects of approvals were still built into the system. However, the simplicity that was introduced into the enterprise user's experience was a key to the project's success and it also had to be learned. None of these features were specified in advance of the execution phase, meaning they had to be learned along the way.

This turned out to be a very successful innovation project. It sold well. Soon after this project, the same SaaS model was developed for HR systems, Supply Chain Management, and Customer Relationship Management which are all part of Oracle's cloud success story as of the time of this writing.

Behaviors

Open Communication: One key behavioral aspect of this team was the emphasis of communication within the team. All of the documents were available in an internal business site. This included documents for product requirements, contact lists for who to speak with on every subtopic, the mission of the project, and information about the user experiences.

Context Leadership including an Industry Analyst: A second behavior to highlight in this case has to do with the organizational structure. Consistent with frameworks for Innovation Engineering, a context leader was placed as head of product management. That product management leader had the benefit of a staff employee on the team who was an analysis. Their job was to additionally act as a domain expert to understand everything there was to know about the application. In this case how do supply chains work, how does the retail industry work, how do firms purchase goods? Of course, there was a corresponding parallel technical leader and technical team.

Summary

This is a sample case for a top-down innovation project led directly from the CEO. Like other company projects, this one follows the model of the start-up within the firm. Since the CEO was the initiator, political coverage and isolation was built into the organizational model. Using a story about the business scope, the execution was developed around a prototype product that continued to improve over time. The team's leadership had very capable people with strong personalities. Learning while executing is also very evident in this sample case.

Chapter 4: The Missing Puzzle Piece

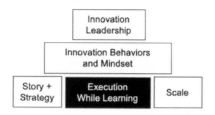

Framework: The Innovation Engineering is actually an inter-woven combination of Technology, Entrepreneurial Management, and Design Thinking.

In this chapter, we introduce a tool called an Innovation Navigator. It is a spreadsheet based tool that can be used to coordinate the entire Innovation Engineering process.

In an earlier chapter, we walked through the steps of the Innovation Engineering process. In the future chapters, we will cover the behaviors and mindset that are needed for successful innovation projects. However, in between these two, there is still a missing puzzle piece. That piece is the understanding of how Innovation Engineering fits within the activities and processes that are already used within any team or organization.

In this section, we clarify that Innovation Engineering is not intended to replace any of the current tools, processes, stage gates, or team activities. Instead, it is a light overlay that coordinates activities, reflection, and communication across the existing technical and business processes and tools that are already in place. We will also introduce the sample tracking tool that we have used for projects. This is the ingredient that any team or organization requires to both learn and execute simultaneously.

We have Observed Many Benefits using Innovation Engineering

We regularly run projects using the Innovation Engineering process, coupled with the tools, leadership model, and innovation culture describe in future chapters of this book.

As we use the approach properly, we have observed that innovation projects run more effectively, multiple times faster, and with greater positivity – as characterized by this representative quote summary:

- "Using this process on this project, we accomplished more over one summer than we were able to in over the last year."

- "The communication within the group improved a lot, there was no need for unnecessary meetings."

- "Onboarding new people used to be very inefficient, but on this project, people were able to contribute from the beginning."

- "The Navigator provides a history so we can look back and see the decisions, accomplishments, and the overall journey."

- "We could use the Navigator to review last week as well as plan for next week. It also makes the team feel good about the progress they have made."

- "I really liked the concept of separating what is working from what is not working."

- "We found that this process was critical in the beginning phase, and then later when it was close to delivery, we could switch to more standard action item lists."

- "The project started unstructured and undefined but using the process, the structure was developed during the execution. We were able to learn what was possible during the project."

- "The story part of the process was worth it because it gave the team a common objective and helped the team come up the learning curve."

- Advisor/Manager Viewpoint: "The results of the project exceeded my expectation. My ideas were in the project, but there was a lot of contribution that was not directed by me. The team was able to bring more of their own creativity and skills into the project deliverable."

Making It Practical with the Innovation Navigator

In this section, we will explain the Innovation Navigator, a project tracking tool, which makes the process concrete and actionable.

The chart as shown is a table version of spreadsheet tool that we use for activity tracking, cross-functional team communication, reflection, and inductive learning.

This version of the tracking sheet is implemented, used in the form of a shared spreadsheet such as a google spreadsheet document. As new team members join the group, they share access to this document.

Feel free to Use our Most Updated Version
This tracking document is available as google spreadsheet document to be copied for anyone's use without any cost at my professional website:

 www.ikhlaqsidhu.com or www.innovation-engineering.co

Innovation Navigator	Tracking	Period 1 (Example)	Period 2
Initial Conditions Team Leads: Context: Technical:	How We Win (Context Lead)	* 3 high quality customers * product able to do X, be very simple to use * freemium model with over 10% conversion	
Concept Summary: e.g. NABC or High Concept	Working/ Known	* technology works * demo is inspiring	
Competition Target:	Not Working/ Unknown	* sales pitch not yet effective * will not scale beyond 40 users/day * user interface not simple enough	
Success Metric:	People to Add	* need advisor to create channel * need help from UI expert	
Context Links: - Link to Story file - Business files -Other repositories	Context Actions Items	* advertise for UI position * informally talk to at least 5 customers this week about the value to them	
Tech Links: - Agile or Dev Files -Messaging group, slack, or other	Top Agile Development List	* Fix UI * find different way to scale users	
	Log 1: Context Lead Reaction 1-5	I am not convinced that the users are willing to pay for this because something similar is available for free at XYZ. Reaction = 4	
	Log 2: Technical Lead Reaction 1-5	The scaling problem can be solved in 2 weeks. The user interface problem requires skills we don't have yet. Reaction = 5	
	Advisor / Management Feedback	Advisor comment: I would solve the UI problem first	
Team:: Sue Xu Bob Smith ..	Scratchpad for Weekly Notes	Scratchpad for Weekly Notes	

Innovation Navigator in a Table Format

Setting Up the Innovation Navigator

There is a portion of the spreadsheet dedicated to the initial conditions of the project. This portion of the sheet is created when the project execution is about to begin. These initial conditions include the following:

- *The names and contact of leads.* Typically, this the context (or business lead) and/or the technical lead depending on who is actually involve in the initial state of the project.
- *The initial concept of the project.* This may be written in NABC format. It might also be written as a high concept pitch. A link should be provided to a longer version of the narrative assuming one has already been developed.
- *Another initial condition of execution is to identify what is being replaced.* This might be a competitive product or company. It might also be another way that people solve their problem now.
- *Up to 3 success metrics* should be listed in bullet form so that a measurable outcome has been set for the project, at least for the current phase.
- In practice, having the links to all project files and related planforms on this sheet is very helpful to everyone on the team. This includes everything from a google group, to scrum lists, to link to a slack channel
- And finally, at the bottom of this column is a placeholder to *add the team members as they join*. These same people will also be shared on this document, so their contact information is available in the share settings of the document as well.

Using the Innovation Navigator

After initial conditions are set up, the only thing left to do is use it. This is done in two ways. First, the leads must update the sheet on a periodic basis. For very new and smaller projects, this may happen on intervals of once a week. In organizations where the project cannot move that quickly, it might be reduced to once every other week, every third week, or even once a month. Second, the cross functional core team should meet once on every interval and review progress using the sheet. This is the best way to achieve alignment between the different functions that people are working on.

At the beginning of each interval, say week 1, the leads would fill in the box on the tracking sheet to capture the current state of the project and make incremental learning and executable plans. See the top 3 rows from the example:

- How We Win (Context Lead): This is actually the question of strategy. Has something changed in the manner of how this project will be successful? From period to period, this row may not change very often. However, there will be times when something new has happened in the background environment and therefore the strategy for the project to be successful will also have an addition or even a significant change.

- Working/Known: This field is also generally filled by the context lead. As shown, it labels those aspects that are no longer open variables or risky concerns.

- Not Working/Unknown: This field in contrast lists those things which must be addressed for success, but as yet are not working or are still open variables.

In traditional execution, the list of action items is created based on what has already been accomplished or from a static list created in the past. In this model, the separation of unknowns and the changes in strategy are the inputs to create new action items. These actions are either experiments to resolve an unknown or they are actions intended to fill a gap of what is not yet working. *This is a key concept: the unknowns and gaps drive the execution.* The execution list is learned, not created from preconceptions of the team.

This process should continue from period to period. By being systematic about the questions to resolve and the actions to take, the project increases its predictability and likelihood of success. The questions being asked force the team to understand the issues, reflect, as well as learn on the path. For naturally innovative teams, this may seem mechanical, but for teams who have not practiced this before, it's a chance to learning an innovation behavior until it becomes natural to them. Furthermore, this process has been shown to increase communication and keep everyone on the project aligned even when the conditions and requirements are changing.

Innovation Engineering: What is Really Going On? And Why it Works

This section explains what makes Innovation Engineering work. We will see that the Innovation Engineering process is an overlay process that does not break other processes that may already exist within the team or organization.

Most existing organizations already do variations of these three things when they consider new projects:

1. Kickoff new projects once they have a business case and matching technical requirements
2. Use Product Management or Project Management processes to guide the requirements over time
3. Develop and release a product, service, or technology

Over time, the world has changed and there have been advancements in each of these areas for early stage innovative projects.

Before	Today
Business case & technical requirements. In past, based on developing a Product Requirement Document and Functional Specification and a stage-gate process to kick of project.	*Story Narrative:* A story that uses entrepreneurial characteristics to explain the user's view and the business case. This story is used to collect stakeholders and then kick-off the project.
Product/Project Management. In past, based on setting requirements.	*Entrepreneurial Product/Project Management* that uses lean methods to achieve a working business model or mission model in accordance with a changing strategy and changing environment.
Product/Service Development. In past, based on a development plan.	*Design Thinking and Agile Development* to put the user's viewpoint in the design and to implement with changing requirements.

The Innovation Engineering Model is intended to build on each of these developments, and yet coordinate the thinking and activities between them. In a general sense, any new project has these components and can be diagrammed in the following manner:

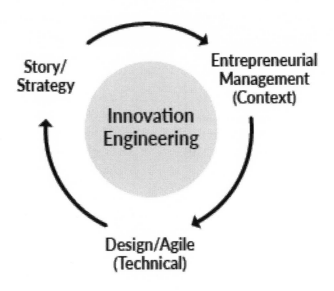

Innovation Engineering is a connecting process and tool to join story-based project kickoff with existing product management and design and delivery processes.

There are multiple options in each of the subcomponents of an early stage project, whether new venture or technical project.

For example, there are multiple types of story narrative and in chapter 9, we cover many of them. The story might also be maintained in a slide presentation or a text document and available to the team for regular updates via a shared disk storage system. And, yet the Innovation Engineering process does not insist on any specific one of them. As long as one version of a story is used to gather stakeholders and kick-off the project, it complies with the intention of the process.

The same is true for the Context or Entrepreneurial Project Management area. There are multiple ways and tools used by a team to keep track customer information, recruit new team members, and get approvals for using resources. The organization can choose to use any of these existing processes and technical platforms to support these activities.

And finally, the same is true for the design and development component. Some teams rely more on design thinking approaches so that the user story more deeply ingrained in the requirements of the product as well as the technical architecture. Some teams will also be able to leverage agile development to a greater degree than other teams depending on the industry. However, regardless of the tools and processes used, the development process must also be synchronized with the learning and reflection that is happening in parallel about the business context. For example, if the business conditions have changed, it is possible that the user's or product features need to change at that instant in the development.

There are two key points to take away from this overview.

- The first is that the Innovation Engineering process coordinates across the components of story, contextual project/product management, and design/development

– regardless of the exact tools, technologies, platforms, and processes that each chooses to use.

- The second is that if needed, anyone can design their own version of the Innovation Engineering process. Any new version still needs to ensure that execution is truly being done while learning and that the initial project requirements are launched from a reasonably well-developed story narrative.

Summary

In this section, we have seen that Innovation Engineering is a connecting process that works with existing models for product management, design thinking, and agile development. It does not replace these, nor does it require a change of tools or software platforms. On the other hand, Innovation Engineering is a coordinating function which starts with story and creates a behavior of learning while executing within any group that is working on an innovation project. In concert, a simple tracking map is used to log the project's status, actions, and reflections. Over time, the behaviors which are introduced and reinforced by the Innovation Navigator become automatic to the group. In this sense, it is also a teaching method for Innovation.

Case Study: VMware Scaling Innovation in Mobile

VMware Scaling Mobile Device Management

VMware, Inc. is a publicly traded software company listed on the NYSE. At the time of this writing, Dell Technologies is a majority share-holder. VMware provides cloud computing and virtualization software and services.[5]

VMware was one of the first commercially successful companies to virtualize the x86 architecture. Its product lines enable users to set up virtual machines on a single physical machine, and use them simultaneously along with the actual machine. More simply, it enables its users to install a virtual operating system within an operating system and use them both at the same time.

VMware founded in 1998 by Diane Greene, Mendel Rosenblum, Scott Devine, Ellen Wang and Edouard Bugnion. Greene and Rosenblum, who are married, first met while at the University of California, Berkeley. Edouard Bugnion remained the chief architect and CTO of VMware until 2005, and went on to found Nuova Systems (now part of Cisco).

This case example starts in 2008 with a concept for VMware to move to mobile virtual platforms. VMware acquired Trango in 2008[6] with the intention to extend VMware's core product to mobile phone platforms.

In this case, the Trango acquisition[7] was a first attempt for the firm to innovate in this area. In a second step, like a pivot, the

---

[5] https://en.wikipedia.org/wiki/VMware
[6] https://www.vmware.com/company/acquisitions/trango.html

firm later completed a second acquisition, AirWatch, which would be needed for the innovative product concept to achieve market scale. In this case, some team members of the original project even joined the new venture.

> *Authors Note: The case shows how an innovation project may start within one team or approach, but then move to scaled execution using an acquisition.*

For context, according to VMware, one set of companies had invested in a particular technology for mobile, but a much larger segment of the industry invested in a different technology. In time, the second technology had significantly more adoption, resulting in VMware making another acquisition in the mobile space. This also resulted in transferring a firm's existing capabilities and competences required for growth into the acquisition during the scaling of the business.

---

[7] According to the company's website, Trango had "developed virtualization software that lets a single mobile phone run several operating systems at the same time." The acquisition announcement also mentioned that "This helps VMware's vision of giving customers secure access to their applications and virtual desktops from any device, anywhere includes mobile phones,"

Use Case

A scaling process allowing growth from $100M to $600M in annual revenue (See Forbes)

2008: VMware acquires a small startup Trango to extend virtualization to mobile devices

2014: VMware acquires AirWatch with 1600 employees (2014 VMware press release)

2019: A Culture change within the group.

Integration: Work to scale engineering and innovation.

"Successfully integrating a 1,600+ person company with 10,000+ customers is complex. The collaboration and partnership, between the AirWatch and VMware leadership teams, made this a success and reality. AirWatch founder and CEO John Marshall worked closely with VMware COO Sanjay Poonen, along with their teams, to ensure the overall business, customer base and product saw massive growth and scaling; in fact, the team successfully created a brand new market category, UEM (Unified Endpoint Management), spanning mobile, Rugged IoT, and Laptops!" - Jayanta K Dey

Initial Conditions:

This case starts originally starts with the acquisition of Trango - to extend VMware's core virtualization capability on a mobile device[8]. In parallel, the industry started to adopt a term called Mobile Device Management (MDM) for this area[9]. Apple is said to have coined the term, and Google Android adopted it to secure mobile devices, including securing enterprise email, enterprise applications, etc.

> *Technical note: MDM uses an approach where applications run in secure containers. This is different from the mobile virtualization approach. Both offer allocation and isolation benefits, but they function differently. Containers virtualize the operating system instead of hardware and therefore containers are more portable and efficient[10]. MDM allowed enterprises to remotely manage their enterprise apps, including new capabilities like the feature to remotely wipe all data in the mobile device that belongs to the enterprise. This is needed if an employee leaves the firm and walk out with valuable information still contained in their own phone.*

---

[8] Other industry players such as ST-Ericsson and ARM were also working to add virtualization on mobile processor and phones [see https://en.wikipedia.org/wiki/Mobile_virtualization]. Soon, many companies, including IBM, Microsoft, Apple, Google entered this space – see https://en.wikipedia.org/wiki/List_of_Mobile_Device_Management_software

[9] See https://en.wikipedia.org/wiki/Mobile_device_management.

[10] https://www.docker.com/resources/what-container

In spite of this difference in the technical approach, the mobile market was growing quickly in 2010-2015. VMware did see promise in the mobile market, and in 2014, they acquired AirWatch, an established firm, which already had a more than 1600 employees and a large sales force.[11] Airwatch was the leading provider of enterprise solutions for Mobile Device Management, Mobile Application Management and Mobile Content Management.

Story:

The VMware team had developed a technology in the context of an evolution for the firm. It was an A X B model of (Virtualization) X (a growing mobile market). This logic was not so different from the famous case of Kodak believing that every new digital photography product needed to still have film as part of the story. While the technical demo was compelling, the neither the business story nor the business organization had developed.

The AirWatch story was the opposite. The business (contextual) story was compelling. They had discovered that enterprise customers wanted to offer mobile phones to their employees with two personalities. One for personal use, and at the same time, a second personality with securely contained professional data.

---

[11]   https://www.forbes.com/sites/rajsabhlok/2014/01/24/what-vmwares-1-54b-airwatch-acquisition-means-for-enterprise-mobility/#4e3ff65a2116, https://www.vmware.com/company/acquisitions/airwatch.html

Coincidentally, the AirWatch product was built in a startup culture, excited to change the world, that built quickly, broke quickly, and as a natural consequence, they were less focused on technical stability and scalability.

> *The AirWatch business story was well developed, but the scalability of the technology was not. And the core VMware organization was well suited to aid in this next phase of growth.*

Execution While Learning:

The efforts continued, with others in VMware joining the AirWatch team to strengthen it, which was led from Atlanta Georgia. The combined team was focused on the growth of the business. The execution model was much like a start-up within a larger firm. The projects directions were customer led. It was all at a very fast pace. And at the same time the engineering infrastructure had to be put in place to scale the business.

Besides new features, security capabilities had to be strengthened and vulnerabilities needed to be closed. The other technical work to accomplish was to make the product scale proof by combining more experienced engineers who were mixed in with more junior, entrepreneurially focused engineers.

Behaviors:

This section is particularly about the behaviors for adding features while scaling the product or service. Scaling means that the product will work not only for 1 user or 10, but also for 100, 1000, 1M or more.

The normal model of testing for scale means that you build and test for 1 user and then 10. The performance characteristics are measured while testing with 10 users. The tests are run again with the load of 100 or 1000 or more users. If the resources don't scale linearly, then something may be wrong. For example, if 100 users take 4% of available bandwidth on a server, then 1000 should take 40%. If it happens to take 60% during the test, then there is something that still needs to be learned about the system.

The issue of scaling from an engineering perspective requires a change in the organization and the product platform. People who have "done it before" know the importance of scale and high quality in the engineering process. The reason is that they know the pain of dealing with the problems that come back when the product is released in a way that does not scale. For example, the Amazon approach to scaling uses incentives of developers to make scale happen with high quality. They require the same developers who developed the feature to fix their own bugs later on. There is a large incentive to build a foundation for the technology that is reliable and basically always works.

Scaling is different than the initial creation. It requires a culture change for the team including an appreciation of an experienced team. New processes, motivations, and leadership behaviors are also requited.

First, silos must slowly be replaced with a common platform. Until the scaling point, the product had been created in organizational silos. Silos were at one time developed to be fast in the development cycle. However, this approach leads to replicated work since each silo tends to build similar technology, but purposed for different reasons. One at a time, the new capabilities required by different teams must be brought into a common platform.

Second, there is a culture change that must occur. In the beginning, the team is driven by sales and by changing the world. Later, the team must be driven by making the product great. And they must also have the discipline to work on the boring, non-attractive problems.

Finally, trust appears as a major motivator in the scaling of the business. The VMware leadership team explained that trust cracks can appear at this stage. The trust symptoms and recommended solutions prescribed by the team follow:

1. A new boss must earn the trust of a team that has already contributed. Under pressure, will the boss take the side of the management and alienate the team, or will he/she side with the team and alienate management? The ability to stay level and non-political is key factor.

2. There is also a trust issue between marketing and engineering in these times. The leadership must be genuine about the one team approach.

Overall, the leadership and team behaviors which best aided the team in this case to scale were to promote a culture of transparency, openness, and collaboration.

Chapter 5: Innovation Leadership

The last section described a tactical process to support innovation. This next section is about the leadership model required for an innovative project to be successful. In this section, we map out the leadership issues, starting with team formation and people selection though the scaling of the project.

Note: While this is not intended to be a book on leadership, it is undeniable that many aspects of team leadership are tightly connected with the process of innovation.

Building the Team

Every team starts somewhere, and that somewhere may just be one individual. This individual is best complemented by someone whose skillset differs from their own. Thus, an ideal starting point consists of two team members: one functioning as a context lead, and the other as a technical lead.

Step 1: The initial team must build and sustain trust

As stated earlier, all teams have one fundamental starting point: trust. When starting with two co-founders, it is crucial that these first two people trust each other before any business begins. Trust applies not only to honesty and integrity but to competence. This means that when one founder says something, their co-founder should be confident that it is correct. This trust in competence applies to all team-members as well. The integrity aspect of trust requires that, given the choice, team-members will make decisions that prioritize the team's interests over their own. These values come first and foremost to any business development; if they are not established at an early stage, the project is unlikely to succeed.

When it is time to recruit a new person into the team, trust again comes into play. A common misconception is that the hiring and recruitment process should be focused entirely on the skill and competence of the recruit. This is only partially true, as skills are of course critical to a high-performing team. However, a more important factor to success is trust: can this potential recruit both trust and be trusted by others on the team? If trust is not maintained, endless energy will be expended to keep information away from that person, resulting in counterproductive strategies to avoid issues around this one team-member. This leads to a crucial principle of trust in the world of business and innovation: **If one single person is not aligned and trusted within the team, the performance of entire team will suffer greatly.** This trust check will apply to every new employee or stakeholder that is brought into the team and network. There is an implicit speed of trust. When trust is high, decision and progress is fast. When trust begins to falter, progress will slow down to a crawl.

Step 2: Consider the balance of team expertise vs innovation culture

Every person on any given team has different skills and capabilities that they bring to the project. Let us imagine that you could rate any given team member on a scale from 1-10 for these major categories:

- **1 Point = Narrow, precise thinker:** This is for a skilled person with *operational or execution-oriented capabilities* including the ability to receive tasks and complete them reliably and with precision. This type tends to be **narrowly focused** on execution and tend to respect authority.

- **10 Points = Broad Innovative Thinker:** This is skilled person with *innovation behaviors* often including communication skills, connector-behaviors, and influence skills as described earlier. This type of person tends to be more **broadly focused.** Often, people in this category may be harder to manage because they don't respect authority. Instead, they think critically and ask for justifications on all assignments.

Everyone is balanced somewhere in between 1 and 10. Some lean towards reliable operations and/or execution of plans, others towards change and innovation.

An important factor is that no team should be overly represented by just one of these skill categories. A team of only high skilled narrow-focused people will solve the wrong problem. A team of all broad-thinking innovator types will perform lots of experiments, but may suffer when it comes to execution. A narrowly focused team has high efficiency but low adaptation and an inability to take significant risks or experimental approaches.

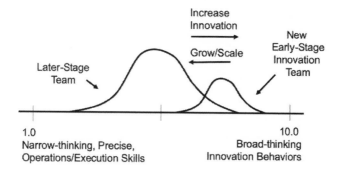

Organizational distribution for Innovation

The goal is to get the right mix on every team. Innovation projects require individuals with better-developed innovation skills, whereas routine operational or execution projects require narrower, technical experts or functionally-oriented people. Operational projects will also require a smaller fraction of people with connectivity skills and innovation behaviors.

The team mix changes over time and the average can be represented as a score between 1 and 10. For an innovation project, the team composition tends to have a greater number of innovation-oriented people in the beginning, averaging somewhere between 6 and 9. This higher average is found mostly in aggressive industries. In contrast, conservative industries often have an innovation culture score on the lower end, even between 1 and 4.

As the project matures and as the number of unknowns in the project are reduced, new people are added. These tend to be people who have been added for their skill and reliability, as opposed to their wider comfort zones, connector capabilities, and innovation behaviors. As the team becomes larger, the balance changes away from innovation skills and towards operational or execution-oriented skills.

For every project, the mix of people should match the objectives. When the project is early, open-ended, and high in risk, then these elements must be matched with the right balance of people. When the project starts to scale, then the mix will evolve along with it.

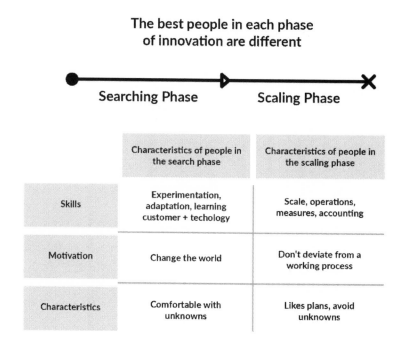

The best people in each phase of innovation are different

Searching Phase   Scaling Phase

	Characteristics of people in the search phase	Characteristics of people in the scaling phase
Skills	Experimentation, adaptation, learning customer + techology	Scale, operations, measures, accounting
Motivation	Change the world	Don't deviate from a working process
Characteristics	Comfortable with unknowns	Likes plans, avoid unknowns

Characteristics of people who innovate vs scale

In some cases, larger and later-stage projects need more innovation and adaptation and fewer operations than they currently employ. In this case, even the larger project must adapt its mix of people to raise its innovation culture score. When this type of correction is needed, there are several options. One is to use coaching to move people farther towards an innovation mindset. Another option is to focus on the innovation behaviors of new hires so that as people leave, their replacements have stronger innovation behaviors. Both of these are active decisions to be made by the management of the project.

Leading an Innovative Team

The creation of an innovative, high-performance team goes through multiple phases, as do the challenges faced by leadership during this time.

Leadership with Trust vs Insecurity

In the early stages of team formation, the main challenge is recruiting and selection. Good leadership requires trust in order to collect team members. It also requires story-telling and social EQ skills. As the team starts to work together, trust continues to be critical in the management of the team. It is important that people feel they will get fairly rewarded for the things that are yet to be created.

The most destructive factor to effective leadership is insecurity. If the team's leadership is insecure, they will start to compete for credit or rewards against the team that they are leading. Many brilliant innovators share the ability to divide credit easily. Rather than boasting about their own accomplishments, they take pride in the accomplishments of others on their team. They always highlight others to make sure that their work is made known and appreciated. If an innovation or insight comes from within the team, they lavish the individual who was the source of that good idea, never stealing the praise for themselves. This is the behavior of an innovative leader, and it is effective for one simple reason: the best people with the most to contribute will not stay loyal to an insecure leader. And if the best people leave the project, the innovation will not be created.

Planning, Urgency, and Agility

As the project, story, and team develop further, the method used to manage the timeline of progress becomes increasingly important. There are two opposing views that need to now be synergized.

On one hand, the team must be able to set a target date for when the project and its milestones will be complete. For example, if the project aims to release a consumer product to be available in time for the holiday season, then completion dates are important. In this case, urgency and focus must be maintained in order to get the product done on time.

On the other hand, agile behaviors allow teams to undergo continuous development in two to three-week intervals. By the third interval, the priorities of the team may have changed significantly from the initial conditions.

These priority will result in parallel changes in the target date of delivery.

Work Backwards with Planning and Work Forwards with Agile Iterations

With two conflicting approaches, how does one decide which should be used? The answer is both. In early story development and scoping, the team must set a major milestone to be reached, even before they know what it will require to get to that milestone. Then, with this date in mind, the team can work backwards. To illustrate by example, consider a product that needs to be in stores by Christmas time. The team creating this product will set their target date of Black Friday as the day after Thanksgiving in US to begin the Christmas Holiday shopping season, and then work backwards to identify several key timeline components: When does the store need it in stock? When must the contract be completed? When will the distributor need the product in-hand? At what point must product testing must be completed?

Having worked backwards, the team now has a starting point, and can now move beyond their story and into execution, using a learning model and with agile techniques. The details of their plan are no longer as important as the timeline and external constraints that they've identified.

The agile process is not meant to predict which features will be available at what time; rather, by establishing milestones, the process provides a sense of urgency which compels team-members to focus their energy on a particular task.

If the team does not use an agile process in the forward direction, timing and urgency may be achieved but it will be applied to the wrong end product, leading the project to miss its target. And if the team does not publicly communicate urgent milestones, the focus and urgency required to achieve progress will not materialize.

| Milestone: | Alpha | Customer Trial | Train Partner | Launch |
| Dates: | 1/15 | 3/1 | 5/1 | 9/1 |

Example: For measurable progress, we first work backwards
with external constraints, then work forward with agile
iterations. Each milestone is a forcing function and is needed to
set urgency around key events in the projects delivery.

The Next Challenge: After Growth Comes Alignment

As the project grows and the team size increases, a new leadership challenge will emerge. In the beginning, the team must develop trust as they search for a working business model. Once the size of the project increases, the issue of developing an innovative approach is no longer as important. The problem is actually having too many projects and too many people pulling in different directions.

This is the challenge of team alignment. While initially there was just one core project that everyone was working towards, there are now 10 variations of that project. With each variation, a different agenda among all the team-members working in their own areas. The new goal is to create a cohesive working environment where everyone is on the same page. The coordination, planning, and direction of the project must be aligned.

This alignment requires strength in the leadership. The leadership should gather input from each team-member on how to proceed, and ultimately develop a plan to work towards the same objective. It is the responsibility of the leaders to communicate a path forward and align the efforts of all members of the team. Therefore, the leader must have a strict framework for how to develop this path:

1. Agree and align all your effort to this one plan/direction
2. Disagree and still align towards this one plan/direction

3. Leave the group, team, or organization.

There are times when large conflicts within the team will be inevitable. Often there is no go-to solution for the immediate challenge ahead. However, no matter how intense the discussions or arguments, the team must finally come together as an aligned group of leaders.

The Innovation Leader's Role

Before this section is concluded, it is important to highlight the true *role* of a leader in the context of an innovation project.

Selling and Influencing

In all projects, resources must be collected. This means that someone will need to convince others to provide the team with resources — the act of selling. It is imperative that a leader spend significant time selling or influencing early on. This act of selling has a dual objective for the leader: The first goal is to acquire resources or and the second goal is to comprehend the customer need or even funding agency's viewpoint. Whoever has the power to make decisions for the business *must* understand the needs of all its stakeholders, thus it is that person's job to generate resources through selling. If this responsibility is passed on to another member of the group, the leader who makes product-related decisions will not have direct information from customers and other stakeholders.

For example, if a customer complains about the product or service and some newly hired team-member is on the receiving end of this complaint, the leader will have reason to blame the subordinate, rather than facing the customer directly. This is bad for business and bad for innovation.

It is only after the project and organization has grown and after the open variables have been greatly reduced that the sales function can be passed to another person trained to perform the job well.

Coaching the Team to Learn Inductively

During the project, the team is trying to execute while learning new things. Underlying this model of execution is the process of "inductive learning." This is illustrated in diagram below. At every interval, the team is looking back at what has worked and what has not worked so far. Using these reflections, they can then build informed experiments directly into the execution path. The Innovation Navigator that we use to track projects was designed to reinforce this model of inductive learning. It allows the team to cycle through reflections and actions related to strategy, team building, technical development, and other activities.

Project Journey and Inductive Learning

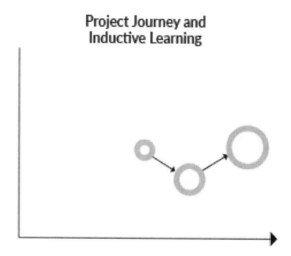

Every innovation project is a journey for the team. Inductive learning is the method of looking at current results from past decisions and asking why those things happened.

The leader also has a sizeable role in this process. In the context of this section, the leader can be considered a "mentor" to the team. If the team is operating independently, they might use an Innovation Navigator spreadsheet to self-organize the project. When a mentor is involved, then he/she needs to provide guidance consistent with the aims of the project. The mentor uses their critical reasoning to see whether there are gaps in the logic of the team. Can their experience provide insight to what the rest of the team has not yet realized?

As a mentor, it is important not to provide the answers to every problem at every interval. Instead, the mentor facilitates the discussion of options. For open questions, the mentor may ask the team how they know whether a certain assumption is correct. Then, the mentor could ask the team to come up with the activity or test which will result in discovery of the correct answer or correct approach. This serves a dual function of teaching team-members to think critically while also identify key issues.

For example, if the team is trying to figure out if users will pay for a specific service, one mentor may volunteer their opinion, however, a better mentor would ask the team how can they answer this question themselves, and eventually lead the team to an experiment to test the preferred selling price.

> ### *Example*
>
> *A photo-editing software company named Blnk is having a group meeting.* Their goal is to figure out whether or not users will pay for their newly offered airbrush editing service. There are two mentors on the team who are hoping to spark some critical thinking discussion. Mentor #1 gives his opinion: he knows the market and he thinks that users will pay for it. Mentor #2 instead asks the team what they think. Then he asks the developers to design an experiment to test the team's hypothesis. **Which mentor do you think did a better job?**

Project Scoping

Another major component of the leader's role is to manage the scope and timing of the project. This component goes beyond mentoring and is closer to the concept of managing the innovation process.

When the project begins, brainstorming often dominates the conversations. A wide range of ideas are floating around about how the project can be successful. In this stage, the project is wide open and undefined. As this process continues, the expansion of ideas and possibilities may become vast and unlimited.

The project leader's job is to continually reduce that scope until a deliverable is achieved. In each stage, and in each iteration cycle, the number of unknown variables should reduce. The leader should apply their judgement and continually reduce scope of the project while focusing on the delivery and timing.

For example, in earlier iterations, the scope may be undefined and many new ideas may have been discussed. The leader's job is to allow all the ideas to flow during the innovation phase, but then to decide which ideas and directions should be pursued and which ones should not.

After another iteration or after another phase of the project, the project scope needs to be further reduced in order to get closer to the delivery. The leader's role is again to first allow remaining considerations, and then to apply judgement to further reduce the scope of the project, making sure to allow a timely delivery of the project.

Self-Development and Team Development

Every person, including the leader themselves, can be characterized as a point on the chart below. The vertical y-axis is strength in a discipline or function such as engineering, finance, or other academic area. The horizontal x-axis represents a person's "Psychology of Innovation." These are behaviors and mindsets described in the previous chapter: a wide comfort zone, low social barriers, and high emotional intelligence (EQ) will all fall upon this axis. Strength in this dimension can be considered street smarts. These individuals typically listen well and are cable of influence others.

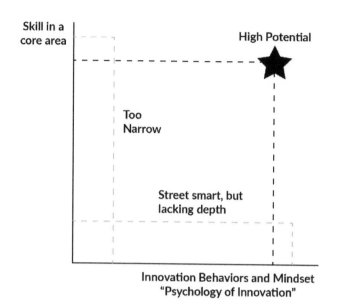

Most training that is done in academic intuitions is designed to increase a person's capability in a skill area on the Y axis. Likewise, in most organizations, most of the people are strong in skill areas on this axis. Some people have a natural and innate level of strength in a "Psychology of Innovation" sense, and fall farther along the X-axis. These individuals who are just marginally farther than others on the X-axis tend to be the ones selected as the managers, directors, vice presidents, and executives. Many also tend to be entrepreneurs themselves. The most brilliant greatest technical innovators of our generation, such as Steve Jobs or Bill gates, are individuals who have high strength in both axes of the graph.

Assessment and Self Development

For a leader who wants to self-develop, the first objective is to identify where the he or she lies on this map. Are they strong in skills and light on the psychology topics? Or perhaps they have strong people/EQ skills, but they need specific skill development? Using this map as a tool provides one with a starting point for identifying a person's capability. By understanding the demands of the position, a leader can decide how they want to self-develop.

Over the years, I have personally overseen the development of many of these types of tools, including the Berkeley Innovation Index, a survey to precisely measure the innovative characteristics in people so that they can both assess themselves as well as self-develop. In addition to this index, see link below, there are also many exercises that can be used to develop these types of behaviors. We invite the reader to search for these tools online.

berkeleyinnovationindex.org
https://innovationindex.berkeley.edu/

Team Development

In addition to self-developing broad and deep skills for innovation, it is important to understand that teams can achieve high performance in innovation by being both technically skilled as well as strong in the psychology of innovation.

This is absolutely key to understand: Each team member does not need to be strong in *both* depth skills as well as innovation behaviors. By grouping together diverse combinations of people, some of which are strong on the Y-axis and others on the X-axis, it is possible to create a high-performing combination that is more fortuitous and successful any one person is likely to be.

The main caveat to this dynamic is that those people with diverse skillsets need to trust each other and work together genuinely. This takes us back to the fundamental behaviors for innovation in the last chapter, where the ability to form trust relationships is a key ingredient. Having an understanding and respect for the things that others can do is the key factor to developing that tight binding of people with different capabilities.

Finally, the leader's job is to facilitate the collaboration of individuals with wide ranging skillsets and diverse capabilities. The leader must embed within these people the ability to form bonds of trust, and to instill in them a common purpose.

The Innovative Leader's Broadest Functions

Finally, and most broadly, the leader cares for the team, making those decisions which benefit others before himself. He who makes decisions to care for himself is not a leader. A leader is someone whose mission is to generate growth, opportunity, and purpose for those who decide to work for them. The leader's scope is to do the right things, not only for their own group, but for the industry as a whole. Championing integrity, this person is not to be viewed just as the leader of a company, but as a leader in society, one who can be looked up to, trusted, and turned towards for advice.

As the company grows, so does the relative power of the leader who runs it. Like a financial holding, the power of the leader is compounded over time, appreciating in value as that person acts on behalf of a wider set of people below.

As is intuitively known, the power of a leadership position comes with the responsibility to work with integrity. Tough decisions are an inevitable part of this process. Whether it's divesting in a subset of the company, laying off an on-profitable fraction of the team, or leaping into a new and unexplored component of business, the leader must always act with calculated, intentional acts, using judgement and intuition to make the call that most aligns with the mission of the company.

Summary

This section is not intended to be an overview of all personal or organizational leadership practices. Rather, these leadership aspects have largely focused on the development and growth of an innovative team. The key factors addressed in this section have   included:

- Using trust to fuel the growth of the team
- Moderating the balance between technical skills and innovative behaviors
- Realizing and implementing the value of connectors
- Achieving balance between planning, urgency, and agility
- Understanding the difficulties and approaches to team alignment
- The leaders role in influencing, mentoring, and project scoping
- The development of the self and team in the pursuit of improvement

As this section ends, we see the can see the multiple roles of the innovative leader including resource gathering by story-telling and influence, project scoping, control of delivery timing, and literally caring for people inside and outside your team.

Chapter 6: Applications for Innovation Engineering

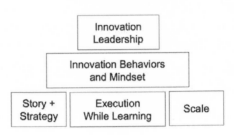

This chapter is intended to provide vignette examples of what innovation problems feel like and to convey why each of these examples is a good application for the Innovation Engineering framework.

Use this chapter to compare which types of innovation issues are the closest to your own situation. Let's start by considering the following project categories:

- Innovation projects developed within firms
- Innovation projects within government organizations
- Innovation projects in Applied Research Labs intended for translated use outside of the lab
- Entrepreneurial projects for new venture creation and acceleration
- Educational and innovative projects as part of degree programs for capstones, master's projects, or even undergraduate team projects focused on innovation.

Each intends to create innovation, each has different modes of failure, and each can benefit from an Innovation Engineering process.

Larger Firms and Government Organizations

These projects usually start with some fundamental problem that needs solved. This solution therefore requires something to be changed, adapted, or built. The project might have internal customers or the project might be a market-oriented product or service.

To fully understand the nature of these projects, we first must be clear on what the projects are not. These projects are not about fixing a bug in something that already works, nor are they about scaling or expanding something that is already implemented as an effective solution. If the firm/organization already knows what to do and there are very few open variables, then it is not an "innovation project."

To further clarify, these are example projects that fall under our definition of innovation:

- A start-up within a larger organization: on occasion a few people are asked to work separately from the mainstream organization to create a new product or service. Usually provided with political cover, they are given the freedom to create advanced features or variations of existing products that might serve a market shift. This may include a new business within a firm like Oracle, Apple, Amazon, or and other larger company. How are these internal ventures originated? What logic is used to develop the story/plan? What is the role of the leadership? What does execution look like in this type of internal venture? How does the model work with an acquisition? There are many factors to this process, besides the more commonly expressed concepts of user validation and development of a minimal viable product.

- A large retailer is concerned about falling behind due to Amazon's disruptive presence. As online sales increase, the retail store's in-house business model begins to falter. They need an innovative solution, and fast. Perhaps this means implement emerging technology, developing an alternative sales approach, or even looking into a company acquisition. However, these solutions raise a wide array of open-ended questions and variables to consider — will customers actually use the new tech? How can this team ensure that the new approach will be successful?

- A government organization fears the potential for fraudulent activity on its website. One day, bad actors hack into the organization's tax department website and then claim to be proper taxpayers. They file a false tax return, collect the funds, and leave the taxpayer and government agency to sort out the resulting mess. Now, some may suggest that implementing AI, data technologies, or a blockchain-based solution would best prevent something like this. However, a broader view may demonstrate that this problem is a result of policy constraints or an outdated computer security system. Either way, the outcome of project is highly unpredictable, and many of these projects will fail in a very costly manner.

- A telecommunications firm is considering new equipment to increase speed and capacity while reducing latency in its network. Before an investment is made, they begin to ask whether the cost of this change will actually outweigh the new revenues that can be generated from it. First, they must determine what new revenue generating services for a telecommunication firm would look like. After generating some hypotheses, they decide to implement a technology and business project. This project may result in a successful new service, but could also result in wasted time and money. It is at least equally possible that the firm is simply not able to develop any services that are competitive with those already offered for free by Internet firms like Google or Facebook.

A first characteristic of these projects is that the path to success can be unpredictable. A great deal must be learned about the technology as well as the context of the problem in order to address it. These are all innovation projects that are run every day. These happen to be real life examples.

In these projects, the major issue is that the project cannot simply be executed by following a list of steps from a pre-determined plan. In fact, the problem is likely to be unclear. Other times, the people involved with the problem don't know how to obtain the resource approvals needed to actually launch the project. When they do launch projects of this type, they tend to fail quite often.

Additional failure modes occur when:

A) The project gets lost in technical details.

B) The team does not agree on how best to proceed.

C) The complexity of the project spirals out of control.

In parallel, the project's funding can be cut at any time. And in the end, the customer or internal stakeholder might not even want the final result.

Applied Research in Universities or Companies

Another application area for innovation engineering is corporate research or its counterpart of applied research in universities. These are both environments where innovation results can be unpredictable.

Innovation in Academia:

We can start this conversation with applied research in university and government labs. There is no shortage of successful technologies and ventures that have come from university research. Take, for example, Berkeley Unix, open source software, biological innovations such as the influenza vaccine, computer-aided circuit design, and even the invention of wet suits; these are all great examples that happen to come from UC Berkeley research.

However, it is still accurate that even the most applied research work typically does not lead to usage in the real world. This partially has to do with the many misconceptions about how and why research gets translated into application or commercialization. A typical problem scenario in a university lab is of this nature:

6. A researcher may have solved a very difficult problem. That problem might have come from an open space in the literature, or it might be directly inspired by a usage case. Often, the research results and/or demonstrations are showcased with industry players and advisory boards. Sometimes connections can be made, but the results are more often left in the labs, while firms move ahead with different approaches and different products. Clearly, the work that can be highly rewarded in academic settings rarely gets rewarded in a market sense, without some intervention or adaptation.

In some cases, an entrepreneur or entrepreneurial professor will see potential in certain aspects of the research. Ironically, the successful product or innovation is rarely something that the researcher intended to develop. Only a few faculty (about 2%) at most institutions have learned the art, behaviors, and social models needed to transform their research into companies. Most have not been able to do it.

One can argue whether university research should be fundamental vs applied. However, in cases where application or new ventures are a desirable outcome, the success rate and predictability have significant potential for improvement.

Innovation in Corporate Research

The case of applied research in advanced development and corporate research labs is even more profound. Development labs primarily focus on building the next product or service of any given company. In these cases, much is already known about the customers and many technical questions have already been answered during the development of previous products.

However, the early stage innovation projects in research and/or advanced development labs have a more complex problem. They typically do not work on the current product line because the development teams are better positioned and better organized to do so. A good example of this is Google, who occasionally works on moonshot projects which are high reward but also very high risk. Most companies cannot afford work on moonshots. In business literature, these are called Horizon 3 projects[12].

As a result, advanced labs tend to work on a few standard categories:

- Future features which are hard to implement or sometimes even hard to conceive. (*i.e.* advanced development of a self-driving technology within a car company)
- Educating the rest of the company about conceptual products that may be possible. (This is often the case at many SV Innovation centers.)
- Or, with the goal of detecting market shifts, the advanced development group tries to analyze new customer patterns, changing behaviors, and product adjacencies that the mainstream company is not able to focus on.

These types of projects also come with their own host of issues. For one, they require a complex learning path with many unknown variables to consider along the way. Second, simply

---

[12] Horizon 1 projects are new product and services for the same customer and same technology. Horizon 2 projects are adjacent products and services that require the learning of a new technology or new market.

being able to demonstrate the technical capability or concept is far removed from success in the market. And third, the ability to execute the next step after ideation or even demonstration is difficult because it requires a complex combination of technical leadership, political support, entrepreneurial behavior skills, and often coordination with business development, mergers and acquisition capabilities.

Some of these groups have been successful in their firms. However, due to the constraints and complexity of their problems, the results in many organizations often stall at the stage of demonstration and strategic planning reports.

Once again, there is potential to increase both predictability and success in these types of early stage innovation projects.

New Venture Creation

Entrepreneurship and new venture development make up yet another category projects that can benefit from the Innovation Engineering framework. This techniques offered in this book are undoubtedly well matched for new ventures. Many successful entrepreneurs already understand the right behaviors, skills, and resources needed to develop new firms. But for those without experience in this area, it can prove to be quite challenging.

The topics in this book apply to entrepreneurship, whether in classroom education, venture acceleration, or as an independent new venture. This is also related to the issue of teaching and mentoring others who are working on creating new ventures. The problem with most classroom entrepreneurship "educational programs" is simple — their results are typically poor. There are, of course, certain world-famous programs known to produce large numbers of entrepreneurial alumni and new ventures. However, these do not make up the majority of educational programs in this area. In fact, in most cases, the program content is purely theoretical and/or artificial in nature. The students struggle due to lack of real experience, often having no genuine contribution to any real project or venture, while instructors struggle because they've rarely (if ever) been involved in an actual entrepreneurial venture themselves. This, sadly, results in a lose-lose situation for both students and their instructors.

Many of these instructors and want-to-be entrepreneurs are somehow under the impression that developing a logical PowerPoint presentation is the same as creating a company. In reality, there is a massive gap between creating slides and creating a company. In between this gap lies the challenge of "execution" and related factors of mindset, behavior, networks, and comfort zone which are crucial within the context of early stage innovation.

Many established accelerators suffer from the same exact issues. There is often a simulated sense of venture creation, when none is actually taking place. The reality is that smart people often spend lots of time and resources, but fail to develop any new successful ventures.

It is interesting to note that graduate students of faculty who have indeed started companies tend to start more new companies as well. This is because the aforementioned factors, such as mindset, behavior, and networks, tend to get transferred along in conjunction with the formal education.

Besides the psychology-based issues mentioned above, a gap also remains for other reasons. Issues often stem from not understanding the process for new venture execution. This is also an area that can be improved to result in higher success and predictability.

The model of innovation outlined in this book is intended to address these gaps and create a workable foundation for venture-technology education.

The Capstone Project of the Future

There are many master's degree programs and even undergraduate programs that offer an innovation-oriented project. The objective in each of these is to integrate the knowledge learned in the degree in a real life in an experiential manner. Quite often the project includes a real-world project with an industry or non-profit organization. The issues with these projects is includes the following:

1. The projects often fail to lead to impact beyond the learning within the team.

2. The company or external interactions are often too low in priority for the firm. This is because any project that is high priority to the firm cannot be easily outsources to a set of students not managed or well-connected within the firm.

3. The timescale between when the project is specified can be many months before the students work on such projects, and in this case, they problem may have become outdated by the time the students select the project. Alternatively, the company contact may have been reassigned or may have even left the company

during this time period.

4. And in default, many of these projects devolve into the job of assisting a professor on a smaller and less applied research area. Or alternatively, working on a very narrow company project to execute a simpler task where innovation is not part of the solution.

Another reason why this lack of innovation happens within these projects is that the project was pre-conceived before the student was ever involved. They don't know the challenge or context that lead to the project and they jump into the project much later.

The fact is that much of the value of any project is in deconstructing the problem into the steps required for execution. In other words, it is valuable to actually discover the problem before navigating the execution of the solution.

Summary

Innovation projects often fail. Innovation projects also occur in many different settings including private firms, new ventures, government organizations, applied research laboratories, and even student projects.

All of these innovation scenarios benefit from the Innovation Engineering framework to make these projects more successful and effective. As we have seen, sample cases have been presented to illustrate the real-life character of these innovation projects within different contexts.

Case Study: Mixbook Navigating Innovation

Mixbook, a start-up created by former Berkeley students Andrew Laffoon and Aryk Grosz, has navigated Innovation for over 15 years.

Mixbook is an online design tool for customizable photos, photobooks, calendars, canvas prints, invitations, and cards. The company, based in Redwood City, California, was founded by former UC Berkeley students Andrew Laffoon and Aryk Grosz. After taking several courses in the Sutardja Center for Entrepreneurship & Technology, Laffoon and Grosz graduated and subsequently started the custom photo-service firm in 2006. Approximately 13 years later, the firm has seen tremendous success.

This case study covers the initial launch of the firm as well as the subsequent launches of its innovative product lines. The behavioral methods, leadership decisions, and strategic paths employed by the company will all be examined, demonstrating how these aspects of the firm fueled its ability to launch, grow, and adapt over the past decade and a half.

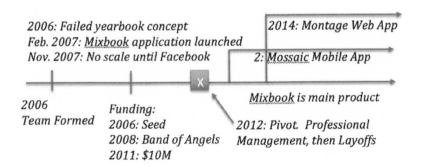

2006: Failed yearbook concept
Feb. 2007: Mixbook application launched
Nov. 2007: No scale until Facebook
2014: Montage Web App
2: Mossaic Mobile App

2006
Team Formed

Funding:
2006: Seed
2008: Band of Angels
2011: $10M

Mixbook is main product
2012: Pivot. Professional Management, then Layoffs

Initial Conditions

The story of Mixbook begins in a class taught by American entrepreneur and investor Jon Burgstone at the UC Berkeley Sutardja Center in 2005. Laffoon and Grosz had taken similar courses and eventually found each other. With a shared interest in starting their own venture, the co-founders continued to discuss their concept with Jon Burgstone even after the course had ended.

> *Andrew Laffoon and Aryk Grosz formed a team after meeting at UC Berkeley with the intention of starting a new venture before they know what the venture would be.*

After some research in 2006, the two were committed to testing a commercial concept that would allow high school students to self-publish their own yearbook with editing software and print-on-demand capabilities. What they hadn't realized is that yearbooks were normally created under the exclusive supervision of a teacher or administrator. Upon meeting their potential customers at high schools, they quickly learned that the schools wouldn't even consider financially supporting this type of product concept. One teacher even went so far as to say that he would "never allow this product at the high school," apparently because it would lead to the loss of his job and would ruin the yearbook since the students were not supervised.

At a loss for how to proceed, the team fortunately had a chance to speak with Joshua Chodniewicz, the founder of Art.com. During their meeting, Joshua saw the drive and expertise of the co-founders in front of him, and decided then that he would seed fund them under the condition that they pivot to the publishing of photobooks. The new idea was to offer online editing software and publishing capability for people to create a coffee-table style photobook through Mixbook. The target audience for the concept was "soccer moms" who could upload family photos and then order their own coffee table Mixbooks.

Story

Because Art.com founder Joshua Chodniewicz expressed interest in seed funding of this concept, the contextual story for Mixbook became the venture pitch slides that were created during the funding process. The technical story was a "paper" version of what would become the future Mixbook website, showing what it would look like when it was actually developed. It was a sort of a low-tech demo. Shortly after the paper demo, the founders developed a live demo. The two founders spent all of their time in this period directly coding a site that matched the paper version of the demo story.

> *Mixbook's story started as a) a venture pitch, b) became a paper demo, and then evolved to a live demo/product.*

The key elements of the initial contextual and technical story were as follows:

1. Mixbook would offer the tools to let its users be creative and collaborative with others to "mix" a photobook

together. (Customizable design, positioning, and style)

2. Due to its collaborative nature, the go-to market strategy for Mixbook would be viral. The team reached out to about 1000 friends or family members and got them to try the product. The hope was users would collaborate with or refer to others, thus creating a network and achieving a critical mass of users through this viral adoption method. (tactics)

> *"The toolset we learned at Berkeley's Sutardja Center were invaluable in getting to an idea that actually worked and then scaling it into a business." – Andrew Laffoon, Co-founder Mixbook*

Execution While Learning

Unfortunately, the viral network effect that they hoped for did not play out as expected; there simply were not enough customers consistently using the online service. The team's hope began to falter, until their market changed dramatically in November of 2007. With the invention of Facebook, user sharing and collaborating became massively popular. Noticing these changes in user behavior, the Mixbook team seized the opportunity and launched their own Facebook application. By mid-2008, the application had reached over 2 Million users on Facebook.

With higher adoption and revenue generation, Mixbook had increased ability to raise funds. In late 2008, Mixbook raised $800K from the Band of Angels, Silicon Valley's oldest angel investment group. With increased legitimacy and user adoption over the next three years, they managed to raise another $10M in funding in 2011, allowing them to scale the firm.

The funding also allowed them the flexibility to make a misstep without sacrificing the core of the business. One of these missteps manifested after the firm decided to hire professional managers to help spur growth. Unfortunately, after observing these managers conduct poor business decisions, such as high spend rates on TV advertising and Groupon, they firm had to conduct a large-scale layoff to bring the business and core team back together.

> *Increased funding allowed the misstep of hiring professional management that was not as careful or knowledgeable about the business as the original founders.*

New innovative product offerings were also launched during this time period. The first was Mosaic, which was a "very simplified" version of the original Mixbook photobook to be created directly through a mobile application. Within just 3 months, this new product was on track to generate over $3 Million/year. Mosaic also found success due to its unique target market — rather than their traditional aim towards soccer moms, Mosaic focused on the older male population.

The next innovation came in 2014 in the form of a high-quality photobook named Montage. Similar to the original Mixbook photobook. Montage also utilized AI tools to allow users to explore more creative design decisions. This newer product, with mobile capability and built-in AI, tapped into a brand new market, found new users, and generated new streams of revenue.

> *While the newer products had found some success, none matched the greater success of the original Mixbook innovation.*

Despite the firm's advances in technology and innovation, the new products did not achieve the same level of success as the original web-based Mixbook application. Having poured resources into these new innovations had to decide which of their three product lines would be most worth pursuing. In the end, the firm decided to return to its core competence and focus on the original Mixbook product. The Mosaic and Montage products would no longer be offered.

Founder Andrew Laffoon gave speculations as to why the next generation of product did not see the same level of success:

1. Too many changes. The rapid development of mobile and AI capabilities meant that there were a variety of open variables left to be considered. It was time-and-resource-consuming to test and validate all these variables.
2. Dilution of focus. Perhaps the issue was that the team spread themselves too thin, trying to accomplish a variety of goals rather than focusing on one at a time.
3. Overconfidence. Having found such large success from the first product, the team may have overestimated their ability to repeat that success with other products.

> *Author's Note: The reader is also asked to speculate whether the core competencies of the firm were leveraged enough in the new*

> *products. This topic is discussed in the story and strategy section of this book.*

Behaviors and Mindset

Upon reflection, Andrew offers the following insights into the behaviors and mindsets that have helped his team navigate the innovation process within Mixbook:

> *The behaviors and culture of the organization developed during the growth of the firm and were critical to its success.*

Customer Focus

"The culture of the organization is to always focus on the customer first. We read stories of how people use our products. We post the anecdotes on our websites. There is a great deal of attention to following the interests of our users. This has also translated to the expectations on how we serve our customers and how we solve the problems that they have."

Hiring: "We like to hire the people who we believe have the most empathy for our customers. We also appreciate a culture of people who are scrappy. They should have grit, perseverance, and a passion to achieve long-term goals. Another key element of our hiring and people development is to develop a culture of caring, because it allows employees to form strong relationships with our customers and with each other."

Transparency & Trust: "From a leadership perspective, we advocate for transparency. We share even the big problems with the team. The foremost goal is to build and maintain trust within the whole group."

Team Alignment: "Another key objective is to achieve alignment between all departments of Mixbook. All decisions are generally OK, but the key is to have commitment and alignment. Leadership at scale simply requires alignment."

Chapter 7: Culture, Mindset, and Behavior

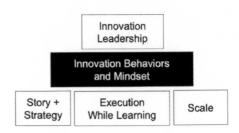

In the last chapters, we presented a tactical process and a leadership approach for innovation that can be used to conceptualize and execute innovation projects. This process, however, cannot succeed without the right mindset behind it. Mindset is an essential ingredient for real innovation, and from one's mindset comes their behavior. When any group of people abides by the same rules and behaviors, developing a shared mindset, we refer to this as culture. In this case, where the objective is innovation, we may refer to it as innovation culture.

More Backstory

At Berkeley, we have had the fortune of hosting a variety of innovative and entrepreneurial guests to come lecture at our university. These lectures provide immense value for both students and faculty, providing insight and generating discussion around emerging industries, technologies, and new ventures. Below is a brief list of these lecturers, a brief "Who's Who" for innovative guest lectures at Cal:

- Ted Hoff, inventor of the microprocessor at Intel
- Mark Andreesen, founder of Netscape and A16Z Ventures, and the 'A' in A16Z
- Ben Horowitz, *the 'Z' in A16Z*
- Diane Greene, founder of VMWare and Google Executive
- Vineet Nayar, *founder of HCL*
- Ion Stoica, founder of Data Bricks
- Jessica Mah, founder of InDinero
- Charlie Giancarlo, Chief Product officer at Cisco
- Marc Tarpenning, co-founder of Tesla
- Michael Marks, former CEO of Flextronics
- Ram Shriram, venture capitalist & early investor in Amazon & Google

In 2010, we noticed that the speakers that we were bringing into our classes offered a great deal more than logical thinking. They also brought with them mindsets, behaviors, and personal rules of engagement for developing effective management, entrepreneurship, and innovation. The observations of these speakers led to our first paper on the topic of the "Berkeley Method of Entrepreneurship" where we tried to pinpoint these behaviors and mindsets, then identify a practical approach to develop these within students.

Around the same time, we began to study the dynamics of an individual's "Comfort Zone," in relation to innovation. One's Comfort Zone is considered a psychological state in which things feel familiar to a person, thus these things come with ease and without stress. Stepping out of this "Comfort Zone" can be high risk, but is often associated with growth and prosperity for the individual. It has been found that this capability to push the boundaries of one's "Comfort Zone" is highly correlated with innovation. In a related body of work known as the "Berkley

Innovation Index," we started to develop the psychological instruments to not only identify, but actually *measure* innovative behaviors in people when possible, allowing us to provide feedback and tools for personal and organizational self-development.

The mindsets and behaviors that are the most relevant to the Innovation Engineering process have been drawn from this collective work, and will be explained next.

Behaviors and Mindsets for Innovation Engineering

Whenever you start an innovation project, the culture of the team — their mindset, behaviors, and engagement rules — are all critical to the success of the project. This section covers the basic cultural characteristics that need to be present within team members. Beginning with broader behaviors which help to create an innovative environment, we will then address those behaviors which are more specific for to architecture, design, development, and technical work.

General Behaviors for Innovation

Story-Telling as a Leadership Model

Many projects are expected to create a "new to the world" type of innovation. Often times the goal is to create a new innovation, even when the terminology necessary to describe the concept might be missing. The innovator or entrepreneur needs to learn how to communicate in a manner which allows potential team members and stakeholders to understand the added value in the new offering. They must learn to communicate their story with a new and powerful language; to do this, it is crucial that they develop their storytelling abilities.

Story is already built into the process of the Innovation Engineering framework. Story is also among the most natural methods of how people learn and communicate, as it reinforces their memory of what they learn. It can be used effectively for several purposes of communication: sparking action, transmitting values, exploring alternative future scenarios, or sharing knowledge. Narration, or story-telling, is central to addressing many of today's key communication and leadership challenges, particularly in regards to innovation.

Trust

Among the most important behaviors for creating any innovation is that the innovator can trust and be trusted. This is important because it affects the speed and quality of information between team members. If a person is considered trustworthy among their team and professional networks, they have the capability of sharing new ideas easily. As they share these ideas, they can also receive honest feedback at a faster rate. This is a mutual learning process. In contrast, when trust cannot be established within a team or any individual, information is held back, and the feedback becomes slow. This stagnates the learning process. It then becomes difficult to innovate fast enough to compete. Of course, just because trust creates faster learning does not mean that every external person should be trusted. Critical thinking is always needed. But after using some judgement, the default option should be to maintain openness so that constructive feedback may continue. By this same logic, team members *must* be selected with the aspect of trust in mind. When trust is on the table, learning improves, creativity spikes, and growth becomes imminent. Trust is critical to innovation.

Comfort Zone

The next most important factor in an innovative team culture is having a wide professional "comfort zone." As noted before, the comfort zone accounts for all that a person is willing to do with ease and without struggle. The action of stepping out of one's comfort zone is paramount to innovation. It shows that a person is willing to try things that are uncomfortable; to develop new objectives and seek out alternative solutions, even when information is scarce.

Before deciding to take on a challenge, a person with a limited comfort zone may ask for training or lots of advance information. In contrast, a person with a wider comfort zone will say "let's try it" or "we can figure it out once we start." They have a confidence that every step does not need to be meticulously planned in advance.

In our work, we have seen that there is a high correlation between wider comfort zones and innovative capability. The now famous "growth mindset," as popularized by Stanford psychology professor Carol Dweck, refers to this very phenomenon. When individuals believe that their abilities can be developed through hard work and dedication, they cut their string of mental limitation and allow themselves room for failure and for growth. The same can be said of this concept; to find success in innovation, a leader must push the boundaries of their own comfort zone and accept the uncertainty that lies ahead.

Accretive Negotiation and Fairness

Negotiation is complicated and everyone has their own approach. One way is to consider the negotiation is as a zero-sum game, meaning that everything a person gives up is a loss to them and a victory to the other negotiator. This is a common behavior that comes with a mindset of scarcity.

Alternatively, negotiations can be approached from a view of what new value can be created as a result of the collaboration; value which neither party had attained before. For example, one party has a customer with a problem, and the other party knows a technical solution. Collaboration between the two will allow for the customer to be connected to the solution, thus creating value for all parties involved. Had either party been close-minded to the possibility of working together, no value would have been extracted. A genuine strive for mutually beneficial solutions or agreements, as supported by a positive sum transactions attitude, is the key to (entrepreneurial) relationships. A "win" for all is ultimately a better long-term solution than only one person getting their way.

As with all negotiations, fairness is an important considerations. If two people work equally to create something new, but one person gets a greater return, then two things may occur: either (a) one of the people will quit in the middle of the project or (b) new people will not want to work with the leader or manager that does not allocate the gains with fairness.

Connectors are key

Connectors are people who tend to have friends and/or colleagues of different types and backgrounds. They have the ability connect people as well as ideas from different places. A connector may know a customer from a social group, a skilled engineer from a professional relationship, and an investor from a business seminar. They can identify connections to be made, and then translate collaborative ideas from one person to another, fitting together different individuals like puzzle pieces.

Many companies have been formed because the founder was a very skilled connector. Innovative leaders actively seek out an ideal team of people and then connect them for projects and for new ventures. In contrast to knowing the problem and solution and execution plan, the connector simply brings together the right people who already possess the skills and knowledge necessary to make things happen.

Diversity equals value

To understand this concept, consider that all people have social barriers. These barriers make us want to talk with the people who are the most like ourselves. However, the people most like ourselves are also the ones with the least value to us. This is because they typically have the same backgrounds, they know similar people, and they may have similar skills. They offer very little new information. However, those who can lower their social barriers to communicate and work with the people who are more diverse and different from themselves end up with increased ability to exchange new information, make new contacts, find new opportunities, and create greater value.

Inductive Learning

Inductive learning is what you might learn from observing the environment and/or conducting experiments. This type of learning differs greatly from watching a lecture or reading a book (yes, even this one). Inductive learning is how children learn not to touch a hot stove. They might have touched it once and then they know not to do it again. Or they observe the effect of someone else touching it. Either way, they notice a result and ask themselves why it turned out that way. In professional environments, this includes the ability to observe what happens in team issues, board meetings, or industry struggles, and develop the ability speculate about the root of the cause. With enough experimentation and validation, these speculations are converted into learning.

Emotional Quotient and Grit

Successful teams also require higher Emotional Quotient (EQ). This means that team members first have the ability be aware of their own emotions, and then the ability to control these emotions. Individuals with high EQ are actually able to positively influence the emotions of people around them. Also related with soft skills such as EQ is the concept of grit. Grit is ability to stick to a task for a long time and persevere through challenges. Grit is also a key element for innovators and innovate leaders.

Summary

In this section, we have looked at mindset and behaviors that are generally conducive of productivity. And in particular, we listed them because they support an environment of innovation. These are mindsets and behaviors that are helpful to everyone in an innovative organization. In the next section, we will look at innovative behaviors that are particularly matched with for technical leaders and architects.

Case Study: Nvidia 2005-2010

A Shift in Technology Strategy and Inflection of a Pioneering Product

Nvidia Corporation, often stylized as NVIDIA, is an American corporation founded in 1993 by Jensen Huang (CEO as of 2019) originally from AMD, Chris Malachowsky, and Curtis Priem, both of whom originally from Sun Microsystems. The firm received $20 million of venture capital funding from Sequoia Capital and other investors.

The three co-founders hypothesized that the next wave of computing would be oriented to graphics processing. The firm produces graphics processing units (GPUs) for the gaming and professional markets, as well as system on a chip units (SoCs) for the mobile computing and automotive market. Nvidia is also now focused on computing for artificial intelligence.[13] By 2019, the firm has approximately 12,000 employees and revenues of approximately $10B.[14]

---

[13] Freund, Karl (November 17, 2016). "NVIDIA Is Not Just Accelerating AI, It Aims To Reshape Computing". Forbes. Retrieved February 11, 2017.
[14] https://en.wikipedia.org/wiki/Nvidia

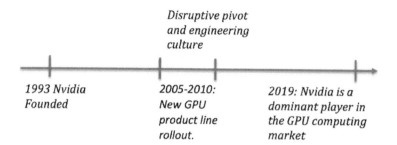

This case describes a project from approximately 2005 to 2010 which created new technology capability during the development of the Nvidia GF100 product line. This capability served as an inflection point for the firm to pull away from its competitors to later lead the industry in broader purpose GPUs and enabled the firm's position to support computationally intensive neural network and AI applications.

One of the significant aspects of this case is about **culture and behaviors within a highly capable engineering organization.**

Initial Conditions:

This case described the R&D process during a sequence of product development releases where a business case had been already developed, however, a new feature capability had been built and integrated into the newly developed Shader Core architectures of the firm, (i.e. Unified Shader or CUDA Core. See a complete list of releases at Wikipedia under the Nvidia related project names CUDA, TESLA, and Fermi).[15]

---

[15] https://en.wikipedia.org/wiki/CUDA#cite_note-1,

While the basic business case had already been determined for this next generation of GPUs, in the background, a change in the industry had been occurring. Nvidia in the 1990s had created the first programmable GPU. This means that a processor (CPU) could request that a function (like filtering an image or adding texture). This function could intern be done by a GPU directly upon the image which would be stored in a memory space shared by both the CPU and GPU. In actuality, the GPU has multiple other levels of functionality, but just as a simplification for those who do not have background, it simply means the Nvidia processor offered a long list of functions that the CPU could ask from it using a very specific set of instructions. This starting point was on the path to a change the entire industry. In the years from 2005 to 2010, Nvidia introduced a new capability under the name CUDA.

---

https://en.wikipedia.org/wiki/Tesla_(microarchitecture),
https://en.wikipedia.org/wiki/Fermi_(microarchitecture)

CUDA: Compute Unified Driver Architecture (From Wikipedia) "CUDA is a parallel computing platform and application programming interface (API) model created by Nvidia. It allows software developers to use a GPU (normally for graphics processing) for general purpose processing. The CUDA platform is designed to work with programming languages such as C, C++, and FORTRAN. This accessibility makes it easier for specialists in parallel programming to use GPU resources, in contrast to prior APIs like Direct3D and OpenGL, which required advanced skills in graphics programming."[16]

CUDA was implemented to work with the "Shader Core", and this was a major shift in product strategy for Nvidia. CUDA's introduction fueled the ability to use a GPU for far varied applications, including even AI applications, and it disrupted its competitors in the GPU space.

From a strategic view and in Clayton Christensen's terminology, this case is both incremental innovation since the product line existed and it is disruptive innovation from a new market sense because it simplified the use of the GPU and made it effective for a wide set of new market applications.

This case starts with the initial conditions that the last generation of non-CUDA processors had been released and that

---

[16] https://en.wikipedia.org/wiki/CUDA#cite_note-1

the development cycle had now began for a disruptive CUDA-enabled product line.

Story:

According to the history at NVIDIA by Forbes[17], the late John Nicholls is credited for taking a key role in developing the use of the GPUs to significantly benefit the High-Performance Compute (HPC) market. Other key members of that team included Ian Buck from Stanford.

GPU computational design requirements are very different from CPU processor design requirements. Traditional CPUs are latency centric. They run one thread as quickly as possible. Whereas GPUs have a massively parallelized architecture designed to maximize throughput as in thousands of instructions in one clock cycle, but are not optimized for any given sequence of instructions to run with a low latency. GPUs in 2019 can run over 10,000 instructions in parallel during one clock cycle.

While the contextual story had been set by the Nvidia business, the technical story was developed through a set of documents, slides, and management oriented pitches by John Nicholls, a principal engineer, who some claim to be the father of CUDA. In this technical story, an architecture showed not only the programmability aspects but also a different type of hardware design, for example requiring L1 Cache for the first time in a GPU.

---

[17] https://www.forbes.com/sites/tiriasresearch/2019/03/29/nvidia-is-a-data-center-company-now/#49a74a086780

Execution While Learning:

As with any advanced R&D project, problems appear during the development of the product that cannot be predicted in advance. In this case, for example in the Fermi project, the product was late in schedule. It also had a problem with the amount of power that the chip consumed. This had to be debugged, understood, and corrected through a series of redesigns.

> *As an example issue, one aspect of the GPU design process is to keep throughput high, which may be at the expense of a higher level of latency than a CPU would offer. However, with some design models, it is possible that the delay can become unreasonably high. One problem occurred with "replays". A memory operation might get scheduled within a finite time, but in not being able to access the memory system, it might get replayed, meaning sent back to the scheduler for later rescheduling. Replay activity might occur repetitively for some memory accesses, thereby unreasonably extending the latency of these operations. (interview)*

Debugging techniques used in these cases were to create histograms of the latency of each thread and look for outliers. The small number of threads that were taking too long would have to be analyzed to better understand errors in the architecture.

In another example, the cache size was increased, but the performance actually decreased, which was the opposite of what anyone would expect.

> *Actually, this occurred in a two-level cache hierarchy. The system employed an L1 and an L2 cache. Between L1 and L2 was a chip crossbar (on chip NOC). Servicing the L1 and L2 were "request buffers," meaning queues that would keep track of requests. Each SM's L1 had a request queue that would track requests made to the L2.*
>
> *What was observed (counterintuitively) is that when the L1 request depth was increased, performance decreased. The root cause identification was discovered after intensive effort on the engineering team's part. Basically, after the L1 request queue size was increased, more requests could be sent to the L2s, and this in turn increased the rate of conflict misses in the L2 caches, causing the caches to eject cache lines that would have hit had there been fewer requests. The L2s basically started thrashing. (interview)*

This was due to having unintended consequences from which data and code was more quickly available to the GPU, and thus changing the order of execution.

The learning lessons from within the Engineering Culture were stated to be:

1. Don't assume anything. Verify all. Or as exempt
   the engineering team, "if it ain't tested, it's broke
2. Measure everything.

A point of interest is that entrepreneurs have been quoted to say the exact same thing but on the topic of developing the business aspects of the firm.

Behaviors, Mindset and Leadership:

Nvidia is known for a highly capable engineering team. Behaviors expressed by team members characterize the culture as follows.

The Nvidia Engineering Culture is characterized by:

1. Hiring the best people
2. Being collaborative
3. Low ego
4. Striving for technical excellence
5. Intellectual honesty, no selling, no making the issue bigger than it is (i.e. blow it up out of proportion)
6. Let data do the talking
7. No which hunts, don't go looking for anyone to blame for technical mistakes

Correspondingly the Leadership and/or Management Culture is as   follows:

1. High talent, very competent, comes from expectations in hiring.
2. The management is also very technical.
3. Reviews are OK, but projects are not micromanaged.
4. Management knows how to get out of the way.

5. Management function is to be like producers of a play or movie. They set the stage and then let the actors do what they need to.

Summary:

Nvidia pulled away ahead in the GPU market, possibly due to a disruptive technology. The key aspects of the story were developed within the R&D organization. The case demonstrates execution and learning in the R&D process. The case also highlights technical behaviors and mindsets required for the team to be successful – and also describes leadership behaviors used within Nvidia to maximize the effectiveness from within the R&D perspective. Finally, the case shows an incremental next generation product, which turns into a new market disruptor. This was the result of a technology strategy developed within the concepts of CUDA and the corresponding evolution of the GPU architecture.

Chapter 8: Technical Behaviors for Innovation

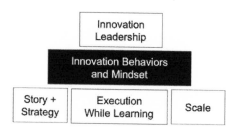

In the last section, we described a set of general behaviors and characteristics that support innovation. In this section, we will focus the behaviors and mindsets that are the most appropriate for technical leaders and technical architects. The text explains how these behaviors intersect with entrepreneurial and innovation processes.

Successful entrepreneurs/innovators have always known how to target the most valuable of problems by using storytelling, inductive learning, and agile implementation processes. In parallel, *great technical architects* have always understood how to solve complex problems by breaking them down to a fundamental systems-level, all while keeping the design as simple and effective as possible.

We have observed that there are significant similarities between the entrepreneurial business process of creating a new venture and the technical process of creating, architecting, designing, and developing an early-stage technological innovation.

System Architecture Development	Entrepreneur/Innovator
• Start with the user's story -> Get to Effective Implementation 1.0	• Start with user's story -> get to Business Model
• Break it down: first principals/relationships	• Use story to gather stakeholders/ resources
• Agile execution	• Inductive learning
• Minimal Viable System: Keep it Simple	• Minimal Viable Product for market fit
• Use broad thinking to reduce failures/flaws[18]	• Use broad thinking to reduce business risk

This table illustrates the parallels between System Development and the Entrepreneurship/Innovation Process

From the table above, we can immediately see corresponding steps between the technical and contextual aspects of an innovation project. With these similarities in mind, we must also acknowledge the many aspects of the process which are specific only to the technical team. In the text below, we will focus the best practices and behaviors that are the most relevant for the technical lead and the technical team.

User-First

Both the technical and business perspective should always start with the user's perspective. Note, this is sometimes considered the customer's perspective. The business objective is to find a

---

[18] Could be written as increase robustness, increase reliability, reduce cost, increase performance (all are true while holding other variables constant)

working business model or mission model, while the technical objective is to get to an effective technical implementation. An important mindset for a technical leader is to begin with the user's viewpoint first. It is only once this viewpoint is clearly established that the system architecture and the implementation can begin development.

Break it down

While the context lead is focused on using the story to gather stakeholders, customers, investors, team members, etc., the technical lead must evaluate potential solutions by breaking the proposed system down into simple sub-systems with minimal inter-connections. Critical thinking is always needed to understand the interactions and causal relationships between subcomponents. Of course, if a sub-component already exists or can be easily obtained, then there is no need to build or redesign that subcomponent. For example, when Tesla created its battery, it created it from thousands of cells that were already being produced in mass scale, instead of designing a completely new battery architecture.

Effectuation

Great technical innovators and entrepreneurs all use "Effectuation Principals" in a natural manner. Roughly, what this means is to start by taking inventory of what you have first. By way of illustration, consider the process of making a home-cooked dinner. Would you first choose an intended dish, and then gather the ingredients (not effectuation)? Or, would you look at what you already have in the kitchen and then invent a new recipe from these ingredients (effectuation)? This principle can be applied to technical and business projects in the same manner.

Look for Insight in the technical story

Search for insight about the location of value, the power, or "the magic" in the system design. What will make it effective or exciting? This concept is a technical parallel to the entrepreneurial behavior of understanding the user's true needs, what they actually care about, or what they are willing to pay for.

Minimal Viable System Architecture

Get as quickly as possible to a 1.0 version. Distill the story as quickly as possible to the simplest possible implementation. From this, a more complex system can be evolved using an agile, iterative model to develop greater capability. This is parallel to the entrepreneurial approach of building a Minimum Viable Product (MVP) for testing product market fit, but in this case the focus is the system architecture for testing technical feasibility.

Agile Increments and Flexible Technology Strategy:

After developing a minimal demonstrable solution, use agile increments to prioritize further development:

1. Start with the simplest possible demonstration on the path to the best solution.
2. Use a technology strategy that allows for easy adaptation.
3. Be agile driven. Accept that it is impossible to predict the final product in advance.

Keep it Simple

The focus of the project should be on keeping the design simple, easy to explain, easy to verify, and easy to debug. Technical architects and designers are often interested in technically brilliant and complex solutions, but true elegance lies in simplicity. As quoted from a historical Apple advertisement, "Simplicity is the ultimate sophistication." You might think of this in parallel to timeless works of art, which are characterized by having exactly what is needed to convey the message but not a single extra stroke of the brush.

Reduce the Downside

Optimize to reduce the downside of risk and failure, not to maximize cost and performance. Always evaluate corner cases. This is the parallel of broad vs. narrow thinking within engineering. The broad thinking version in business would be used to avoid business risk as well as predict the expected outcome in the broadest terms.

Measurable Objectives

Develop measurable objectives to identify when goals are being achieved; you cannot improve what you cannot measure. For example, in a data science algorithm, how will you know that the prediction is good enough? Having both a measure and a target allows you to estimate whether the work needed to achieve a marginally higher result is even worth the expense of doing that extra work. To understand this more, learn about the concept of Value of Perfect Information. See Hubbard, "How to Measure Anything", Wiley.

Create a support ecosystem

Build a support ecosystem with the highest quality partners that you can both reach and trust. Many technical leaders are tempted to reach out to the more convenient lower-quality contacts (team members, suppliers, partners, customers, etc.). These may be easiest to contact, but will not result in the highest quality outcome. As long as trust can still be maintained, it is best-practice to push out of our comfort zones and find the most ideal people and organizations that we can.

Summary

This section was written to provide a wide-ranging understanding of the behaviors and mindsets needed for technical teams and their leaders to successfully develop innovative projects. In case examples, we will see that these same behaviors come up frequently. We also note that behaviors beyond the ones listed have been identified in these cases.

Case Study: Imprint Energy and Christine Ho

From Research to Company to Behaviors

Imprint Energy is a high technology start-up company co-founded by CEO Christine Ho and Brooks Kincaid in November 2010. The firm invented and has been commercializing a breakthrough battery technology that can be literally printed resulting in a low cost, ultra-thin, flexible, rechargeable battery technology.[19]

Since the creation of the commercial venture, the firm's products have been highlighted by MIT as one of the top 50 technologies in the world. The technology development was initially funded through from research grants, then later SBIR government grants, then eventually professional investors. The investors as of 2019 have included Phoenix Venture Partners, Semtech, u.life fund (GVIP), AME Ventures, In-Q-Tel, and Flex Lab IX.[20]

This case covers the Imprint Energy's example of learning while executing. This includes a) learnings related to innovation of the technology and product, and b) learning related to developing the business.

---

[19] See https://www.crunchbase.com/organization/imprint-energy#section-overview
[20] Per official company website.

169

Another topic exemplified in this case is one aspect of the personal development that occurs during a transition from a researcher's role to an executive role.

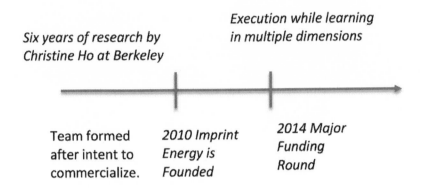

Six years of research by Christine Ho at Berkeley

Execution while learning in multiple dimensions

Team formed after intent to commercialize.

2010 Imprint Energy is Founded

2014 Major Funding Round

Initial Conditions:

The roots of the company come from Christine's 10 years at Berkeley studying material science. Her Ph.D. dissertation topic focused on developing very thin batteries that could both a) be printed by an ink-jet style printer as well as b) be environmentally safe.

The initial conditions of the firm came from Christine's research and a belief that this research work may have value[21] as a commercial product. After leveraging several campus resources including those at the Sutardja Center, she met her co-founder, who was a Berkeley Haas MBA student at the time.

> *In this case, the research work came first, then the intent to commercialize, then the founding team formation, and then finally the development of the venture.*

Story:

Since many years of research had already gone into the project, many early forms of demonstration as well as technical papers were already available to the team. The team searched for both a product strategy as well as a method to fund the firm. The preliminary story of the firm included the following:

a) The decision that they would target or enable new applications for some part of the battery market.

b) That they could demonstrate technical feasibility with sample technology demonstrations and research

---

[21] Note: The author had worked with Christine Ho during the time that she was in the Ph.D. program at UC Berkeley and while she was evaluating the business opportunity. Christine and Ikhlaq Sidhu are co-authors of a study on strategies to clear IP from UC Berkeley for the purpose of technology commercialization.

Execution While Learning:

A focus of this case is what the people at the firm learned through the evolution of the firm's creation and growth. Prior to the first round of investment, the goal of the company was to show that the technology itself would work.

The firm also learned to earn money even before the technology was mature. For example, they found that some firms would pay between $10K USD and $50,000 USD for technical reports or contracted work needed to evaluate the technology for their own future planning. Many firms had strategic interest to develop new battery solutions leveraging Imprint's unique technology and ability to rapidly prototype. At the same time, Imprint Energy's resources were scarce and therefore the firm simply could not afford to do this extra work without using significant resources and expending opportunity cost.

Imprint Energy was in a position to demonstrate great technical capability and discuss product options with prospective partners and customers. Because Imprint Energy had significant expertise in the area, the more serious customers and partners found that it was more effective to contract Imprint Energy for NRE (via non-recurring engineering costs) to run tests and/or write the technical reports that would be needed in the next step of commercialization.

A second stage of self-learning occurred after the first major funding round. In this case, the firm's investors required that Imprint be able to show that at least 1 or 2 early customers had tangible interest in proceeding with Imprint on a project or product development and would be willing to pay $100K-$1M to fund a deliverable of some type. The learning process was to understand how to get to this larger scale for projects conducted at the firm.

Beyond the size of the project, the firm was also asked to show the following:

1. That customers will want a battery based on this type of technology
2. That the product would eventually become a standardized form so that the process and technology could be used at scale.
3. And that a pipeline of additional customers could be developed.

In this phase, after demonstrating technology to potential customers at tradeshows like the Consumer Electronics Show (CES) in Las Vegas, the firm found that the value of a green (environmentally friendly) battery was a major selling point.

Behaviors Learned for Developing the Business:

One unfortunate pattern that developed during the development of the business was that a prospective customer would come to Imprint to discuss possible advanced projects. But after investing time with the customer, the project would suddenly end. The customer might simply stop communicating or they might be pulled back to their company for more mainstream work.

The team discovered the pattern that if the customer contact was not a senior person within the organization, then the project was unlikely to move forward. In the cases, where a CXO of the firm showed interest and spent time with the firm, then the projects were much more likely to go ahead.

In a related example with a large prospective customer, one of the executives said that the firm would simply not offer products that needed Imprint's technology. However, at another point, the team pushed to have this discussion with the CTO of the firm, and then later the project did actually move forward.

> *According to Christine Ho, "Figuring out strategies to get early validation from top decision makers at companies was learned, it was not instinctual."*
>
> *"We also learned that we didn't have to agree to explore or test everything suggested by customers just to satisfy their curiosity, but rather needed them to justify the project effort from a strategic business standpoint. This had to be unlearned from our training as researchers within an academic institution"*

One learned behavior was that if the customer was a technical scout or even medium level manager in the firm, Imprint's executive team had to learn to ask them to bring a higher level person back for the next meeting. Either the higher level person would get involved or they would literally not hear back from the mid-level person. Either way, it would save time in the process.

The executive team also learned that customers have both personal and professional interests when they evaluate any opportunity. Personal interests may lead to a promotion or even personal satisfaction in the knowledge gained by the project, but the business does not necessary develop in this case. They found that it was important to test for genuine business interests.

There are of course many other leadership and entrepreneurial behavior that researchers develop as the lead commercialization efforts. This case only highlights a few examples.

Summary:

Imprint Energy is a high technology firm leading the industry into new battery capabilities. The firm has been entering a new market with a leading-edge technology. This has required market selection, business model selection, and ecosystem development. And in parallel, the founding team has been transforming from an academic orientation to an executive or commercial orientation. In some ways issues of the firm are parallel to the famous E-Ink 2005 case written by Harvard Business School. However, in Imprint's case, they have navigated the challenges more successfully and much more efficiently.

Chapter 9: Developing a Better Story

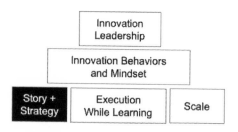

So far in this journey, we have discussed the problems and solutions to innovation. We have provided a step-by-step recipe for addressing and maintaining innovation in early-stage projects. We then analyzed the most fundamental individual and organizational behaviors needed to curate an innovative environment.

Now we will dive into the story narrative of your project. The next two chapters will cover:

- which type of story model fits your project best,
- how to develop and perfect that story model,
- and how to avoid common mistakes made by many entrepreneurs.

Story Development

This section will cover the various models used for effective storytelling and will show you how to choose which type of story to use. There are two types of organizations for which we will choose and curate an individualized story:

(1) Non-profit or public service organizations with a mission
(2) For-profit companies and private firms hoping to generate revenue

In both types of organizations, heavy focus will be placed on common logical and strategic errors to avoid during story generation.

We began this discussion in Chapter 2 and 3 on the topic of the contextual story and its importance. As previously noted, there are several standard models used to develop contextual stories. Among the best models for verbal story telling are the **high-concept pitch,** the **elevator pitch**, and the **NABC story model**. These models are succinct and effective in explaining the purpose of your business or project in a concise manner. Over time, these verbal models can then be developed into a venture pitch, which typically lasts about 20 minutes and can be presented in slide format. Depending on the goal of your project or organization, try to determine which one may be right for you.

High Concept Pitch

The high concept, first introduced in Chapter 2, relies on an "A intersect B" model of explanation. The term "high concept" originated in movie production, where it was commonly used to denote a movie pitch that was a combination of two others:

Examples:
- The movie Jaws, but set in Space = *Aliens*
- Uber for food delivery = *Postmates*

- Amazon for the Chinese market = *Alibaba*

The high concept model is easy to explain, easy to comprehend, and is most efficient when (A) is used as an existing model and (B) as used as a new concept or new market for that model.

Elevator Pitch

Then there is the proverbial elevator pitch. The elevator pitch has long been praised for its ability to provide a succinct and persuasive sales pitch for any project in a short period of time. Communicated in two minutes or less, the elevator pitch gets its name from the time needed to ride an elevator with the CEO from the bottom to the top.[22] The elevator pitch may also contain the high concept pitch within it.

Example Elevator Pitch (Tesla)

Why does Tesla exist? We have record high CO_2 levels in the atmosphere resulting in steadily increasing temperature. And, it's still climbing. Combustion cars emit toxic gasses too, killing 53,000 people per year. What can we do to change this? How can we make a difference? What we're trying to do with Tesla is accelerate the world's transition to sustainable transport? At Tesla, we make great electric cars. This is really important for the future of the world.

A sample Elevator Pitch by Elon Musk about Tesla highlighting a problem and solution while getting people's attention

---

[22] https://en.wikipedia.org/wiki/Elevator_pitch

The NABC Model

The NABC Model is an acronym for Need, Approach, Benefit, and Competition. This model was developed at SRI. The following excerpt[23] from SRI explains the intention of the model:

> "All proposals and business plans must, at a minimum, answer these four questions: need, approach, benefits, and competition - the fundamentals that define a project's value proposition. The NABC approach described here helps us focus on answering these four questions. It is the first step in creating a more complete proposal or business plan.
>
> Example:
> I understand that you are hungry (the need). Let's go to the SRI Cafe (the approach). It is close, the food is tasty, and the atmosphere is quiet there so we can continue working (the benefits). The alternative is McDonald's, which is unhealthy and noisy at lunchtime (the competition or alternative)."[24]

The NABC model is intended to be developed quickly and communicated repetitively at regular team meetings over the course of the project. This helps maintain alignment between the business team and its developers.

The Customer Story

The Customer Story is like a commercial for the product or service. Common formats include:

---

[23]   https://nielschrist.wordpress.com/2012/07/13/the-nabc-method-standford-research-institute-sri/

[24] SRI International Best Practice,

- TV or print commercial
- Product announcement
- Customer case study: problem, solution, result
- Often, it is simply a product demonstration or mockup

Most customer stories demonstrate how the user will use the product and often highlight the more exciting features of the product or service. Customer stories of this type aim to hit the nerves and emotions of the viewer, making the story more memorable.

The Future Press Release

The future press release is a fictitious one-page document that shows what a news story would look like after the successful completion of a current project. There are many examples of future press releases online. Linked below are templates for these press releases along with instructions for how to best create your own:

- https://www.pressreleasetemplates.net/preview/Press_ Release_Outline
- https://www.wikihow.com/Sample/Press-Release-for-Fashion-Show

The 6 Page Plan by Amazon

Next, consider the 6-page plan made famous at Amazon.com. "A lot has been written recently about how Jeff Bezos and his executives run their meetings. Amazon has made it very public that their senior team (the S Team) has banned PowerPoint, and that each meeting starts with a six-page narrative memo which the executives read and absorb for up to 30 minutes before the conversation starts."[25]

> *Full sentences are harder to write, [Bezos] says. They have verbs. The paragraphs have topic sentences. There is no way to write a six-page, narratively structured memo and not have clear thinking.*[26]

Amazon executives call these six-page memos "narratives." These can resolve conflicting opinions in projects that are intended to lead to happy customer outcomes and innovations. Each narrative has four main elements:

"The six-page narratives are structured like a dissertation defense:

1. The context or question
2. Approaches to answer the question – by whom, by which method, and their conclusions
3. How is your attempt at answering the question different or the same from previous approaches
4. Now what? – that is, what's in it for the customer, the company, and how does the answer to the question enable innovation on behalf of the customer?"

This explanation was provided via Amazon: How are the six-page "narratives" structured in Jeff Bezos' S-Team meetings? – Quora[27]

---

[25] Adam Lashinksy, Fortune.com columnist
[26] See "Amazon's Jeff Bezos: The ultimate disrupter" – Fortune Management
[27]https://www.anecdote.com/2018/05/amazons-six-page-narrative-structure/ And see this additional reference, http://blog.idonethis.com/jeff-bezos-self-discipline-writing/

Amazon's narrative style plan has common elements that can be applied to a proposal as well. The first portions of these narratives begin with context and situation and then move on to analysis and activities proposed, generally ending with a budget and total cost estimate.

The Venture Pitch and Its Variations

Finally, we arrive at the most notorious of all story narratives — the investor and venture capital pitch. This type of presentation is typically accompanied by 12 well-developed slides. These slides, known as the Dirty Dozen in some circles, are meant to complement every aspect of the pitch. The structure of this pitch can be applied to a variety of other presentations, including corporate projects, executive presentations, new ventures, government initiatives, etc.

As you will see from many examples on the internet, the most common Venture Pitch outlines look something like this:

1. The problem to be solved
2. The solution proposed
3. The value proposition for both users and customers
4. The market size and opportunity for growth
5. A description of the actual product — how it works and what it looks like
6. The sales, distribution, and revenue models
7. The uniqueness of the technology — often referred to as the unique selling proposition
8. The competitive landscape
9. The team and their complementary strengths
10. The financial projections, including funding needs and financing strategy
11. The timeline of the project from beginning to present

12. The "ask" of the pitch — what is the team "asking" their audience to do?

The slides do not need to be in this exact order. However, the sequence of slides on problem, solution, market, and business model usually come early in the order of the presentation. If the team has a great track record, then sales and previous achievements may come first, but if less experienced, the team is better off placing those slides towards the end.

For examples of venture pitches, many are available here:

https://piktochart.com/blog/startup-pitch-decks-what-you-can-learn/

Which Story Type Should You Use

In the Innovation Engineering framework, we do not advocate for one singular story type over another. Every team differs in the goals and needs of their organizations. Therefore, most teams are better off starting with a simpler model like a high concept pitch, an elevator pitch, or an NABC pitch. These are likely to be the first version of the team's story. These pitches also benefit in that they can be communicated in 2 minutes or less.

Once the team has gathered initial feedback on their pitch, a more complex presentation may be developed. This presentation may follow the venture pitch model or the Amazon style 6-page proposal. To complement this, a customer story can be used as a touchpoint for customers, while the PR News release format can be shared with the team to establish a vision for the entire organization. In this manner, it is important to understand and consider every story type as necessary for the organization or company being created. Ultimately, leadership should decide which narrative type(s) will be most effective to gather stakeholders and achieve alignment in the project.

Revenue Generating vs. Non-Revenue Generating Projects

Many projects are internally focused, and as such, they may not have as many outside customers. There are also cases where the product or service is offered to external customers but revenue is not generated directly from that product or feature. For example, at Apple, many of the successful projects are measured not by revenue, but instead by how often those features are used by Apple's customers.

In the case of a non-profit or public service effort, success may be measured by how the project achieves its mission. For example, the mission of a government organization may be to provide efficient and high-quality administrative services to the public. Success, in this case, is not measured by revenue or income, but by measurable improvements in customer satisfaction, speed of resolution, and overall efficiency. Even though these are not products or services that are intended to be sold, their story is still entirely relevant.

Example
Consider a security project intended to stop data theft within a government organization. Without any financial components, the relevant story slides can still include the following adaptation from the venture pitch version:

1. Problem (or Need)
2. Solution (or Approach)
3. Who is it for? (clarification of customer/market)
4. Value Proposition and/or Benefit

5. Mission Model (what is the organization's mission, and how does the entire project clearly support that mission)
6. Competition, positioning, and other ways to solve this problem today.
7. The team behind the solution
8. Financial Expectations — cost of changes or improvements (sometimes needed)

The contextual story may generally be very similar to the story for a new venture, but with one main exception — the business model is generally left out and replaced with an explanation of how the organization's mission will be served by this project.

Perfecting the Story

The first draft of your story may have gaps, holes, missing information, too much information, a poor solution, or any other variety of issues. This is ok. In fact, this is normal. The story does not need to be perfect. The main goal is that it is good enough to get started. To achieve a better story will take time, energy, and lots of practice, but it all begins with one basic story to get started.

The story will naturally adapt and evolve over time. This process is normal since the story is told and retold to different stakeholders. The best approach is to start telling project's intention and story narrative to credible stakeholders. By listening, collecting feedback, and realigning, the story will naturally adapt into the one that makes sense for everyone involved.

Judging the Story

As your story evolves, it is important to be able to judge whether the project is worth investing time and resources into. Based on the Berkeley Method approach that we have used over many years, we tend to judge early stage contextual stories by measuring three identifying elements:

1) Excitement: The project story narrative must be compelling. If people tend to get excited about the project's story, its excitement will be rated high. Rate Excitement from 1-5.

2) Logic: The story should be logical, meaning that overall it makes practical sense. Rate Logic from 1-5.

3) Social: The team must demonstrate that they have enough social power to attract stakeholders and partners. Rate Social from 1-5.

If you are in the position to compare multiple stories, it is recommended that you multiply together the scores of Excitement x Logic x Social and use the product to measure the story. This number can be as small as 1 and as high as 125. Of course you can normalize to a range of 1-100 by dividing the result by 125. This is a very simplified way to measure the potential of any project or team at a very early stage - so simple that it can be done on the back of a napkin

With each project option, consider this rating system to compare and identify the top projects from any set of candidate projects. This measure is based on the **BMVS model** was developed as part of the Berkeley Innovation Index (BII).

The BII is a research project developed under my guidance at the Sutardja Center of Entrepreneurship & Technology. Berkeley Innovation Index (BII) is a more advanced and more complete set of tools for measuring project potential and is publicly available to anyone without cost. See this tool and others at

https://innovationindex.berkeley.edu/
www.berkeleyinnovationindex.org/

This approach has been tested at Berkeley with many venture pitches and projects. So far, the results of BMVS have turned out to be virtually identical to the estimated results of venture capitalists and seasoned executives when both are used to estimate the potential of a project and story combination.

Now you have all the necessary tools for selecting, curating, and perfecting your story narrative. Remember to practice consistently, allowing for your story to adapt and evolve as new stakeholders are introduced. Next, we will look at the most common story telling errors and strategic mistakes that innovators tend to make.

.

Case Study: Netscape and E-Commerce

An objective to change the Internet E-commerce Industry

Netscape is a brand name associated with the development of the Netscape Web Browser, originally created by the company Mosaic Communications Corporation. The company was founded on April 4, 1994 in the midst of peaking public interest of the Web. Netscape was the brainchild of entrepreneur and computer scientist James Clark, who had recruited Marc Andreessen as his co-founder and Kleiner Perkins Caufield & Byers as initial investors. On August 9, 1995, the firm behind Netscape exploded in value with an extremely successful IPO. The Netscape stock traded from 1995 until 1999, when the company was acquired by AOL in a pooling-of-interests transaction ultimately worth US $10 billion. Today, the firm is part of Verizon Media which later acquired AOL.

For background, it is helpful to understand that team who created Netscape are among the people most credited with creating the Internet as we know it today. Netscape is also credited with the creation of the JavaScript programming language, the most widely used language for client-side scripting of web pages. The company is also known for developing SSL, the first cryptographic protocol which provided unprecedented security in online communications.

The case of Netscape illustrates how an existing firm can take innovative strides forward and push into an unexplored and strategic realm of creation, using insights from their learning processes intertwined with its execution process.

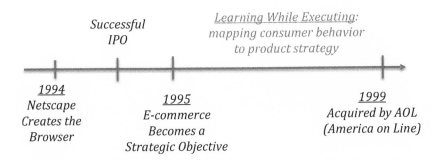

Initial Conditions

The initial team members associated with this project included Marc Andreessen, Ben Horowitz, and Mike Homer. Netscape notoriously recruited team members from Apple and Silicon Graphics, both quality firms with a heavy focus on user interface as well technical leadership.

Netscape firm had already reached a significant level of success given an IPO in 1995. The browser succeeded in making the Internet accessible to millions of people across the globe. Despite widespread adoption, Netscape did not directly benefit financially from this product; the firm offered their browser software by download at no cost. At this point, Netscape still needed a business model!

Story

The Netscape team hypothesized that they would need to participate in electronic shopping if they were to have a sustainable business model that would generate revenue. Their plan was to offer a secure browser interface and back end for consumers to purchase goods and services online. While today this is not a significant concern, at the time it was a large problem for consumers; the idea of entering a credit card number into a browser created by someone you'd *never met* was an idea that most people would not entertain.

> **High Level Story and Strategy Together:**
>
> *The Netscape browser still did not have a revenue generating business model. The strategy of the firm was to participate in e-commerce by creating the secure infrastructure that would enable credit card payments over the Internet.*

Far from perfect, the story narrative championed at Netscape still had a number of flaws and gaps. Their story was more of a strategy that described how the project might be successful. The business context of the story was nebulous at best, not clarifying *how they would extract value* from internet shoppers. It was also not derived from a user's perspective or need. Despite these shortcomings, the idea of a secure payment interface began to form the starting point for technical development and the business activities within the firm.

Netscape proceeded to launch many technical projects backing the creation of infrastructure that would allow shopping to one day be done over the Internet. These projects included:

- Security servers with the ability to hold certificates for trust.
- Browser code which would connect securely with web servers.
- SSL, the cryptographic security standard.
- Netscape, the web server itself.

Execution While Learning

The famously tech-driven firm was now pursuing its e-commerce strategy. However, lacking a true understanding of the user's perspective, Netscape still had several barriers that needed to be overcome. Most notably the issue of consumer safety; people felt uncomfortable typing credit card numbers into a web browser. To learn more, Netscape designed and executed various surveys to learn online consumer behaviors. Individuals were asked to compare, in A/B testing format, their preferences among the following 4 options:

1. A red URL to mean unsecure, and a green URL to mean secure.
2. The use of an 's' at the end of "http" in the URL (https://) to signify security
3. A padlock icon on the browser to show that the site was secure
4. A security certificate displaying the trust-ability of the website.

And then people were asked the key question: "Which of these would make you feel secure enough to enter your credit card number at that web site?"

> *While in the learning phase, the Netscape team realized that brand name was a bigger factor in overcoming security concerns of consumers. This insight changed the product strategy and feature prioritization.*

The answers written in the forms were flatly disappointing. Consumers basically said that they did not care no option stood out amongst the rest, and no option would really make a difference in their online purchasing. At the time, they simply were not be comfortable with online use of a personal credit card; it was a bigger issue than colors, icons, or certificates. Had the team only considered the quantitative survey results, they would have developed one marginally better feature, but the industry would not have developed.

Key insights began to arise when designers had open-ended dialogue with the consumers. They asked "what would actually make you comfortable enough to enter your credit card?" To their surprise, the answer turned out to be the **brand name of the firm.** If the firm was a company that consumers knew and had a relationship with, they would not mind entering a personal credit card number. Through identifying user behaviors, Netscape learned that using a familiar brand was paramount to a user's trust.

This was a **key operational insight**. The team could actually use this information to change their entire product strategy. With this newfound knowledge, Netscape decided to only issue trust certificates to those who had reputable brands. Over time and with continued adoption, the padlock icon and use of "https" became features that signified a secure and trustworthy brand.

Behaviors

In this case, the business aspect of the story was never well developed. Microsoft eventually replaced Netscape's' bowser with Internet Explorer. We should note that strategy did bring anti-trust lawsuits to itself in the process.

> *The case of Netscape case is only a partial success. While they did not develop a successful business model, their innovation did lay the groundwork for Amazon, Google, Facebook, and the entirety of the Internet to develop as we know it today.*

Despite Netscape's inability to stay competitive, their idea of executing based on user behaviors set a very positive precedent that is still maintained today. The technical innovations that came stemmed from Netscape now account for our ability to maintain massive online networks of communication, financial management tools, and an abundance of market opportunities. Netscape achieved this through combining the technically driven teams with an understanding of the user's behaviors, and then incorporating these behaviors into the features of all Netscape products.

Additional behaviors highlighted by the team include:

User Validation: First, the team broke the model of going straight from the idea to code. This means that user validation and user perspective were given a higher priority in the process of development. The team learned that skipping user validation due to time pressure was not beneficial in the longer run. This was at a time when the design thinking model and cycle of prototypes and reflections had not yet become common in technology development.

Context Management with Domain Experts: In a model parallel to the Innovation Engineering process, Netscape's leadership had a great framework of development. Program managers to prioritize the feature list, sophisticated technical architects with knowledge of the system's interactions, and project managers who were in charge of the entire operation. A key differentiator was the role of Domain experts who were assigned the task of understanding the context of the shopping and retail industries. The context leadership job was divided by the product manager and the domain expert. Despite these operational advantages, the team was still unable to create a successful business model or story-narrative. From Netscape one can learn the importance of understanding the user's story as well as the technical design story.

Summary

Netscape started with just strategy and only a partial story. They then executed their product while learning user behaviors. Although they did not manifest into a financially successful business, their innovations paved the way for innumerable technical advances over the past several decades, and will go down as the foundational framework which shaped the Internet as we know it today.

Chapter 10: Common Strategic Errors and Story Narrative Mistakes

What could possibly go wrong?

As the story evolves into a full 20 min slide presentation, it will be subject to evaluation, skepticism, and scrutiny. In order to pass the stage gate[28] and potentially reach agreements with investors or partners, the team will have to practice due diligence, anticipating all challenges, objections, and questions that may be thrown at them.

Newer projects must place significant emphasis on the viability of the financial model. When the project is being conducted within an existing organization, the story place more emphasis on the relationship to the company's core competencies.

---

[28] A Stage gate is part of a process where management agrees to further fund the project allowing it continue on to its next phase.

To be clear, the development of the story is at most 5% of the effort of the project. The rest lies in execution, which is much more critical.

This initial story will inevitably evolve and transform, meaning that in the beginning, it is a waste of time to try and make the story perfect. That said, there are a number of mistakes that should not be made in the story and its accompanying strategy. The sooner these mistakes are corrected, the fewer resources will be wasted in the long run. This section explains these mistakes.

All Logic, No Emotion

The story usually offers problem and solution. An easily made but powerful error occurs when that the story is conveyed without emotion. This error is characterized by the lack communication of actual pain and frustration to a particular customer persona in describing the problem, or a lack of joy and excitement when outlining the benefit of the solution. Ever find your mind begin to wander when a lecturer, manager, or professor drones on about a story in a monotonous and bland tone? This is human nature. People want to be excited, inspired, and called to action. If a story packs no emotional punch, the listener is left with a feeling of "who cares?" Therefore, leaving out any glimpse of emotion, humor, or human relatability will result in a story that is not memorable. It will not stick, it will not attract others, and it will not inspire a call to action in its audience. In short it, will not work.

The best stories are genuine, relatable, and emotional. In any story:
a) Logic is valuable and pertinent.

b) Credentials are necessary with many audiences.

c) But **emotional connection** carries the greatest weight.

This section will be short because the basic point that has been made is simple. However, the reader should note this point. Don't forget it. This point can be the deciding factor in receiving an investment, landing a partnership, or hiring a top-of-the-line employee. Do not underestimate the power of emotional connection.

Solving the Wrong Problem

A very common mistake is that the story is centered on a solution to a non-existent or insignificant problem. This generally happens when no one on the team has spent enough time studying or working with customers. A situation like this arises in one of in two ways:

a) Version 1: The problem being solved by the team is not actually a problem for the customer. The team wants to believe badly that they have the skills or capability to solve this problem, yet in reality, the team has assumed a problem that does not exist.

   A simple example: An innovator has expert knowledge in advanced statistics and its mathematical applications. He or she believes that others would benefit from having this knowledge and would pay for professional advice in the subject. The target customers don't really have a desire to learn statistics, and certainly wouldn't pay a premium for these complicated mathematical applications that are rarely used in day-to-day life. The innovator tries to convince the customer that they need this knowledge, and when the customer does not show interest, they discount the customer's objection. The problem here is that the innovator cannot *hear* the customer. The voice of his or her own beliefs (i.e., ego and self-assurance that they know more than others) is so strong that they are not able to understand the customer's actual viewpoint.

b) Version 2: A genuine problem exists, but the proposed solution does not actually solve it. This means the problem is validated and the solution is not.

5. **A simple example:** The customer says, "I do not need to know advanced mathematics, but I do need someone to estimate my tax payments for next year." The innovator then says, "No problem, I can teach you advanced statistical analysis and then you will be able to estimate your taxes for next year." The customer's problem is now known and validated: they have trouble estimating their taxes on their own. Yet the proposed solution is not what the customer asked for. Even if the customer does agree to pay for the proposed solution (elaborate statistical knowledge), they will not likely feel that their problem was solved as quickly or easily as it could have been. The solution was in the area of the problem, but did not directly address it. In this sense, it was not a validated solution. Again, the innovator is not listening; the innovator is instead offering what he or she *wants* to offer, not what the customer *wants* to receive. Being smart and having a large ego are two factors that often get in the way of the innovator or entrepreneur. Having awareness of

oneself, setting the ego aside, and actually *listening* to the customer will demonstrate empathy and benefit both parties involved.

Unquantifiable Value Proposition

A Value Proposition is a method of explaining the value of a project to investors, managers, or customers. It is commonly written in a soft, qualitative manner:

Example:
We help/offer to **X** (who)
to get/have/receive **Y** (what)
by creating/doing **Z** (how)

This might be: "We help college students (**X**) get better jobs (**Y**) by teaching them how to program in Python (**Z**)." This simple idea is far better than many projects that don't start with any value proposition. That said, while it may be a good start, it does not suffice as an entire plan.

A much better approach would be to have a **quantifiable value proposition**. The idea is identify the cost of the service or product you provide and then to compare that to the next best alternative option (in the case that your project did not exist).

As illustrated in the diagram below, a horizontal line has been drawn. At one end is the mark of cost required to produce the product or service. At the other end is the cost of the next best alternative. When the innovator creates the product or service, a quantifiable value is captured, represented by the length of the line. When the value created is large, the project has some room for errors and learning. When the value created is small, there is little to no room for error.

By offering the product or service at a price in between those two points, some value is captured (selling price of product minus cost of product) and some value is given back to the market or user (price of the next best alternative minus the price of this new product or service).

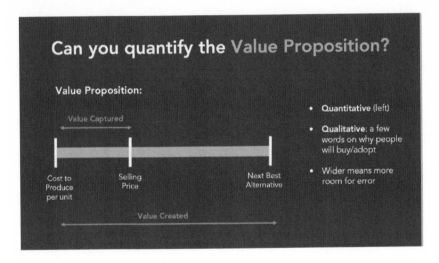

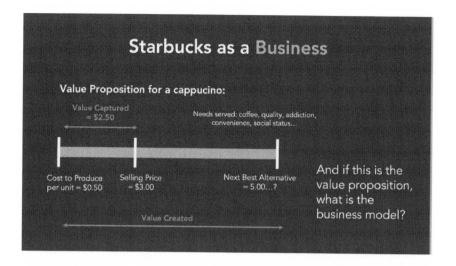

Not all value propositions are this straightforward, but nevertheless the quantifiable approach always works. In some cases, a free service, such as Google's search engine, is provided. In Google's case, the value created for users is real because it is inherently valuable and free to use. However, the value proposition diagram would not be made for the user in mind, it would be made with the advertiser in mind. Google created a new way to advertise, identified next best alternative, and took the difference. This can be diagrammed to demonstrate the value created.

In the Starbuck's example, the reader might argue that a cheaper cup of coffee is available if made at home. However, that is not a fair comparison. The Starbuck's value proposition includes not just the coffee itself, but a combination of quality, convenience, and affiliation with a trendy social experience. A designer handbag (Gucci, Louis Vuitton, etc.) or a Rolex watch is not priced by the function of the product. There are many cheaper bags and watches. The value proposition must be compared with the alternative designer product of similar quality to identify the actual value that the product is providing for its customers. In the case of Rolex, the function is to illustrate the customer's membership of an exclusive social club, and to convince others that they are successful and have specific aesthetic tastes. Alternatives that demonstrate these customer values are products or material belongings of equal or greater monetary value. An elaborate and grandiose house, diamond-studded jewelry, or the purchase of an extravagant yacht provide the customer with similar status that a Rolex does. In any case imaginable, the quantifiable value proposition diagram always works.

The problem with non-quantifiable value propositions is that they don't take competition into consideration, thus demonstrating a lack of aptitude and monetary consideration in the team doing the proposal. This is a mistake that can easily be avoided in the development of the story narrative.

Market Size Nonsense

Quite often, the story needs to communicate that there are many people who will benefit from the product, service, or project in general. This is usually done by communicating a market size. All too often this is done in a very poor manner, and in those cases, a great deal of story credibility can be lost.

The first mistake is to communicate the wrong market. This is much more common than it may initially seem.
Example: An innovator is making a new type battery for electric vehicles that holds 2x the power capacity of the alternative and is priced the same. Here are a few wrong ways to do this:

- **Common Mistake #1:** Find a market research report for the size of the electric vehicle market.
  No: The problem is that electric vehicle market is not the same as the electric vehicle battery market. Innovator loses credibility.
- **Common Mistake #2:** Look up the size of the electric vehicle "battery" market.
  No: This also does not work because once you introduce your new battery, the number of batteries needed per car drops by ½ due to power density.

The right way to do this is to estimate bottom up. How many people will buy an electric vehicle in next year (cars sold/year)? If your battery is adopted, then how many batteries will be needed per car (batteries/car)? Finally, what will be the price per battery (cost/year)?

Total Addressable Market = (cars sold/year) x (price)
= (cars sold/year) x (batteries/car) x (price/battery)

It is not advisable to start with a market research report because there are many built-in assumptions that are unrelated to the innovation being created. A bottom up analysis is always better, as it shows independent research and arrives at a more accurate answer.

A great example of market size calculation can be found in the Airbnb market sizing pitch deck. This calculation was taken from PitchDeckCoach.com and reproduced here below:

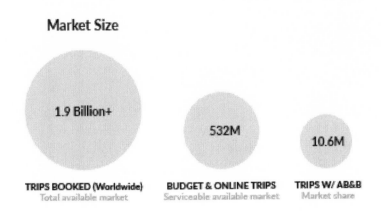

In Airbnb's case, the Total Addressable Market (TAM) is the number of trips taken worldwide every year. This number is reduced to a Serviceable Available Market (SAM) based on those trips that are budget-conscious and booked solely online. Lastly, the number reduced to the target number that the firm thinks it can sell to in the first year. This is the SAM multiplied by 0.02 or 2%, the amount of the market they believe they can penetrate. The resulting calculation is 10.6 Million. By simple math, if Airbnb makes $20/per trip booking, they can estimate their target market to be 10.6 million x $20/per trip = $212M in that first year.

Finally, let's quickly cover why anyone cares about market size. There are two very simple reasons. (1) To ensure the market for the product or service is big enough that you won't run out of customers before you recover the investment. 2) To show that you know how many units you intend to sell in the first year of availability. Everything else is academic detail.

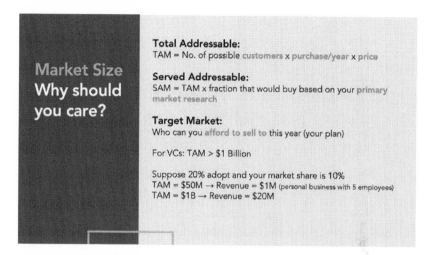

In summary, calculate the market bottom up, not with a market research report. Make sure to get the right market (if you make spark plugs, don't quote the entire auto market). Finally, use the wider market figures get down to the **target market** that you will actually try to sell over the next year.

Financial Model Errors

Whenever products and services have revenue generation associated with them, a financial model is needed. Here we introduce this concept as a placeholder and list a few common mistakes:

Just because people will pay for a service or product does *not* imply that the financial model is sound. The basis of a financial model is the Customer Acquisition Cost (CAC) and the Lifetime Customer Value (LCV):

    a. **CAC** is the amount per customer that one has to spend on advertising and selling in order to secure that customer. This value is often estimated with early sales.

    b. **LTV** is the prediction of net profit attributed to a customer over the entire future relationship between the seller and the customer.

    c. Hint: LTV>>CAC

    d. If it cannot be shown that LTV>>CAC, then the financial model is entirely broken.

One caveat is that many of today's products and services are offered through freemium models. This means that some product or service is offered for free, and then the customer has the option to pay for more services. In this case, the innovator should consider the cost of offering free services as part of the CAC. This cost is no different than an advertising cost designed to attract a paying customer. The LTV can still be calculated in the same way. For most cloud software businesses for example, LTV/CAC should be greater than a factor of 3, otherwise the business will simply take too long to achieve profitability.

For more understanding of financial modeling, we encourage the reader to study managerial accounting, a specialized branch of accounting intended for management to make internal decisions. This is different from studying GAAP accounting, which is intended to specify how financial information should be reported to external stakeholders like the government and tax agencies.

Core Competency Errors

This last section is a placeholder for one to consider core competencies as part of the project's strategy. Projects within existing organizations will benefit most from this section. When a new initiative is started in an existing organization, there is always an evaluation of whether the project will build on the core competencies of the organization. If the project does not, then there is no reason to initiate the project within said organization.

Introduced here is a brief and simple **Core Competency Framework** that has worked for many companies in real life. In the example below, a few of the firm's core competencies are listed in the left column. These are selected based on the historical competitive advantages of the company.

Core Competency	Ability to Scale 10X	Difficulty for others to Copy	Required by Project Option A	Required by Project Option B	Required by Project Option C
Customer Service	medium	high	high	medium	low
Logistics	high	high	low	high	low
Technology	high	medium	medium	medium	medium
Supply Chain	high	high	low	high	low

For each competitive advantage, an assessment is made based on several core components: (a) Can the firm maintain their advantage if they scale to become 10X bigger? (b) How difficult is it for others to copy this advantage? If both values are estimated to be high, then there is indeed a strong competitive advantage within the firm. Components can be annotated as low, medium, and high, or by a simple 1 to 5 scale rating. However the annotation, high values show stronger core competencies.

For every project consideration, the innovation team must ask if each core competency on the list is needed for the potential new project to succeed. If the strong competencies are also needed for the new project (as illustrated with project option B in the table) then that project would be one that fits well within the existing firm.

A Final Note on Getting New Projects Started within any Existing Organization

Although it is presented last in this chapter, it is definitely among the more valuable concepts to understand for any team proposing their own project within a large organization.

This mistake is one made by many teams. It is a crucial consideration, because when the team does not understand this issue, the project will never receive funding and will most likely fail. The question that a management team will ask is two-fold: "should this project be done within this firm?" and, "is this group the right set of people to work on it?" If the latter is not answered early on, the project has miniscule chances of ever achieving success.

The Wrong Way: To propose a story-based project that is too far from the expertise and capabilities of the team or group that wants to develop it. If the project is too far away from the team's skillset, the management will say "This is a great project, but you are not the right team to do it. We should find a real expert in this area, or acquire these skills in some other way."

The Right Way: The team should communicate a story that is in close proximity to the things that they already do well, but stretch just a little toward the direction of the vision that they would like to reach. If the proposal is close enough to the team's skillset and tells a compelling story, and if an incremental project that can show results in about 100 days (this number varies by industry), then the proposal will get a great deal of positive feedback from the management. With this approach, the management will likely respond by giving the team more resources, and then slowly allow them to expand their own core competencies. In larger organizations, incremental works very well!

Appendix of Additional Case Studies

Case Study: Starting an Agile Implementation for Technical Delivery

While Agile is a well understood process, this short case example is provided to understand how to begin the process. A first version of an Agile Implementation is equivalent to having a version 1.0 of the technical story.

In our work at UC Berkeley, we have used this outline below with success, often referring to it as a "Low-Tech Demo". This version of a Low-Tech Demo a technical story and communicated within a 5-10 page slide presentation:

1. How the project functions and what its purpose is.
2. User's perspective: consider the top 3 user expectations including user interface examples
3. Key technical components along with levels of risk
4. An architecture of the system to be created
5. Short-term plan and assignments towards the simplest demonstration.

As mentioned previously, the Agile process is covered in many sources, however the illustration below still serves as a helpful visual tool to understand how we launch an agile process in our programs at Berkeley.

The team that is formed may be aware of a high-level business/contextual story, but the technical implementation is not yet clear. Together, the team should brainstorm possible technical solutions in an open-ended manner. Then, in a norming phase, they should reduce their ideas to generate a low-tech demo as outlined above.

Next, the team designs the simplest possible demonstrable system. The team must be able to demonstrate a minimal viable demonstration within 2 weeks, no matter how simple or skeletal. After this, the Learning Journey may inform the agile process to select and reprioritize features and technical approaches.

Note that many behavioral and cultural norms are used to increase innovation speed, as we have written before. In a 13-week project, there is time for at least 4 agile sprints of about 2-weeks in length after the demonstration of the pre-MVP. MVP is shorthand for Minimum Viable Product, which is the simplest product that would satisfy the users. Our goal here is to demonstrate a concept that will lead to an MVP through iterative agile development.

Initial Conditions for an Agile Developement Process

This process may vary based on organizational context within firms as well as project length and project complexity. However, they key aspects and method of launch stays the same, as does the level of integration with the Learning Journey.

Case Study: Data-X, A Precursor to Innovation Engineering

Fostering the entrepreneurial process in data science innovations

Data-X is a course at UC Berkeley developed and led by Ikhlaq Sidhu and Alex Fred Ojala launched in 2016. The Data-X class is an example of an applied data science course that was specifically designed to address the big picture needs of industry as it relates to the emergence of AI in different verticals, while also adopting lessons learnt from the Berkeley Method of Entrepreneurship (BMoE).

The BMoE principles, which form the foundation of the Data-X program, stress the development of the right mindset for incorporating innovation with technology, which is not accounted for in most entrepreneurship academic curricula. Later, concepts from this program were incorporated in the Innovation Engineering framework.

Data-X Addresses Was Originally Designed to Address these Fundamental Problems:

1. A lack of applied or real-world experience in teaching data science
2. A lack of focus on implementation and powerful tools
3. A lack of integration between theory and practice
4. Correcting the mindset and behaviors required for innovative, technical projects

Data-X was a Precursor to Innovation Engineering:

In designing the Data-X class, we hypothesized that data science applications are very similar to entrepreneurship. There are no specific rule based methods enlisted in a book that can be used to derive value from data. The idea is bigger than teaching data science alone. Creating solutions and transformation of processes in the real world using data science are the innovation problems. To engender innovation in addition to technical skills, we need to train mindsets and behaviors of the individuals working on data science problems. We believe experiential training is the key additive ingredient in training mindsets and behavior's.

The Data-X class ties to the ideas of BMoE which qualifies inductive teaching methodology to be more successful in teaching subjects that involve skill and mindset. Mindset, tactics and infrastructure are the three main components encompassed by BMoE. This method uses experiential teaching tactics that help students understand general concepts through practice, observation, critical thinking, and games.

This method has been used in UC Berkeley for teaching entrepreneurship from the past three years and produced thousands of students as innovation leaders in developing entrepreneurial startups or innovation in big companies. In Data-X we have attempted to use the BMoE method to teach innovation using Data Science.

Data-X Methodology

The figure below summarizes the data-x model, note the two simultaneous paths running throughout the course, one for learning concepts and tools and the other is the collaborative application of concepts and tools on projects. The contents cover math theory and computer science tools, yet at the same time, it includes a project which is developed from an insightful story and uses agile methods to create a solution. The idea is to make students aware of the state of art methods and tools used in the real world and let them work in teams, think about problems in hand with different perspectives and create an insightful story to arrive at a solution.

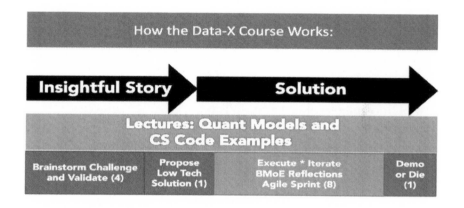

The Data-X Framework shows that story is developed prior to implementation in this early version of Innovation Engineering

Data-X Results

The result of combining story in the form of a low tech demo along with theory and an execution while learning style including agile sprints have been effective. Data-X represents a model which is an early form of the Innovation Engineering framework

The key results in the first two years have shown the following:

217

1. Delivery time of complex projects in shortened timescales of 13 weeks
2. Demand of the course increased exponentially within 2 years
3. Demonstrable applicability to job placement
4. Development of innovation mindset and behavior within a technical course

Project Case Examples:
Data-X projects are publicly available and posted at the site:

See data-x.blog

The case examples that follow demonstrate a combination of story and agile implementation in a period of 12 weeks.

Data-X Case Example A: maia: AI for music creation

By Edward | Yichang | Louis | Cale | Frank | Yiran | Andy

Join maia on her musical journey of self-discovery and development. Experience the maia's creativity and diversity as she continues to mature into a professional composer. Browse her latest works and discover the practical and theoretical unpinning that powers maia.

Our story

We started out with the intention of creating an AI that could complete Mozart's unfinished composition Lacrimosa—the eighth sequence of the Requiem—which was written only till the eighth bar at the time of his passing.

We ended up creating a deep neural network called maia—play on the term 'music AI'—that can generate original piano solo compositions by learning patterns of harmony, rhythm, and style from a corpus of classical music by composers like Mozart, Chopin, and Bach.

We approached this problem by framing music generation as a language modeling problem. The idea is to encode midi files into a vocabulary of tokens and have the neural network predict the next token in a sequence from thousands of midi-files.

See more at https://edwardtky.wixsite.com/maia

Data-X Case Example B: Stayfe

UC Berkeley students create app that can map your safest route home

By Bilal Halabi, Seung Woo Son, Ki Hyun Won and Avery Yip

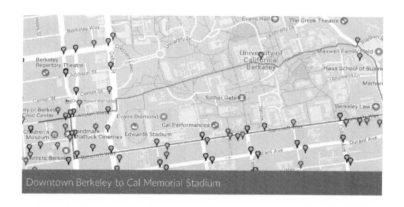

Downtown Berkeley to Cal Memorial Stadium

At times, living in Berkeley can feel unsafe. Shootings, violent demonstrations and robberies are just a few among many incidents that have threatened Berkeley students living on campus this year.

To address this problem, a team of four UC Berkeley students who have experienced these threats first-hand set out create an app to help make residents in the area safer. The result is the web-app platform, Stayfe, which allows users to visualize and track in real-time where crime is occurring in their neighborhood and beyond.

Currently, a medium of this type does not exist in Berkeley. Alerts from Berkeley Police Department or the University of California Police Department can sometimes come hours or days late. But Stayfe uses a customized crawler, parser and crime type classifier to search through news articles in the Bay Area to get the most up to date information on local crime.

Additionally, the students created an algorithm for the app that can use this crime data to suggest the safest path for you take to your destination if you're walking. This feature gives users an alternative to Google Maps, which may lead pedestrians on a quicker, but more dangerous path.

Though the creation of the app was challenging, the students say that their passion and belief in their mission helped them to create a successful product and led them to be selected as finalists at the 2017 SCET Collider Cup.

Moving forward, the team plans to deploy their web app as a mobile app, so people can use Stayfe's safety path suggestion feature on the go.

UC Berkeley students create app that can map your safest route home
https://data-x.blog/projects/map-safest-route-home

Data-X Case Example C: Prism for Tracking Influencers in Social Media

One of the biggest problems that advertisers face is finding the best social media influencers to promote their brands. In this industry, which is projected to hit 10 billion dollars by 2020, finding the right influencer is important. To tackle this problem, five UC Berkeley undergraduate students set out to help companies find influencers to help their brands by building PrISM (Predicting Influence in Social Media).

PrISM is a product that gives companies advanced analytics about influencers on Twitter. In addition to basic statistics such as number of favorites/retweets, PrISM provides a much more in depth analysis of an influencer's profile to help brands reach their target audience. These advanced metrics include follower count over time and activity over time. This allows brands to see how an influencer grows and gives some perspective about how where the influencer is trending for the future. PrISM also uses different machine learning algorithms and techniques to get a follower demographic breakdown, run sentiment analysis, and perform topic modeling on a user's tweets.

A demographic breakdown allows brands to see what ages and genders an influencer is reaching. Sentiment analysis enables brands to see how well an influencer reaches an audience. Topic modeling allows brands to see what topics an influencer likes to tweet about, so they can align themselves with appropriate influencers. After compiling these different analytics, we displayed these statistics and graphs in a beautiful, customizable card dashboard. Early reactions to PrISM show that it has a lot of potential and could prove very valuable to advertisers.

See http://github.com/AdeelCheema/PrISM

Epilogue

The purpose of this book, even in this first version, has been to explain how innovation actually works and to provide a framework to execute the innovation. We introduced this framework as the Innovation Engineering. The tactical process was designed to bring inductive learning into the execution of any project. At the same time, it has been designed to reinforce the behaviors and mindsets that are also needed for innovation projects to succeed.

The reader may find it of interest to know that even this book's design and construction was completed in just a few months using the very same Innovation Engineering process. A screen shot of the spreadsheet tool that was used to "Navigate the Innovation" of this book itself is provide in the illustration below

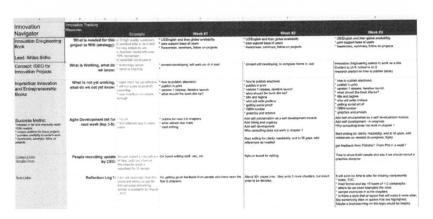

Now, as mentioned, this is only a version 1.0 of the text. Future versions of this text will include additional cases examples, more

information specific to execution within new ventures, and discussion of research-based technology transfer.

This first version will soon be augmented with additional content and cases after we learn (while executing) how teams and leaders benefit the most from the principles within. Thank you for reading, and good luck with your innovations!

Index of Term

Accretive Negotiation
The practice of negotiating in a gradual manner to build trust and satisfy both parties of the negotiation.

Agile Development
A method of software development based on iterative progress, where requirements and solutions evolve through collaboration between cross-functional teams.

Agile Implementation
The process of executing a flexible project plan based on developing small portions of a project at a time.

Alignment
A state of agreement and cooperation among a group about the common goal or viewpoint underlying their organization.

Berkeley Innovation Index
A research concept developed at UC Berkeley which offers simple but powerful ways to measure an individual's innovative mindset and ability to innovate. Found here:

www.berkeleyinnovationindex.org/
https://innovationindex.berkeley.edu/

Berkeley Method of Entrepreneurship
A unique method of entrepreneurial training in which students learn inductively and in a journey based context by identifying problems and working backward to develop proper solutions.

Comfort Zone
A behavioral space in which activities and behaviors fit a routine and pattern to minimize stress and risk. This zone provides a state of mental security. Stepping out of this zone is risky, but results in higher growth and greater opportunity

Connectors
People who tend to have a wide network of colleagues, friends, and associates from diverse backgrounds. These individuals are uniquely skilled at connecting other people to spark collaborative ideas.

Core Competency
A defining capability or advantage that distinguishes an enterprise from its competitors.

Customer Validation
The act of validating the efficacy of a sales process and the assumptions that underpin a business model.

Data-X
A framework designed at UC Berkeley for learning and applying AI, data science, and emerging technologies.

Effectuation
Refers to thinking, based upon scientific research, with the desire to improve the state of the world and enable individuals to create firms, products, services, markets, and ideas.

Emotional Intelligence
One's capacity to be aware of, control, and express their emotions, and to handle interpersonal relationships judiciously and empathetically.

Emotional Quotient (EQ)
A measurement of one's emotional intelligence, often represented by a score in a standardized test similar to IQ.

Execution While Learning
An alternative method of execution in which business proposals are implemented, tested, and learned from before they are considered validated or complete.

Fixed Mindset
The belief that one's basic qualities, like intelligence and talent, are fixed traits. These types spend more time documenting their intelligence than developing it.

Flexible Technology Strategies
A strategy or solution which can be adjusted easily to an organization's specific needs without significant effort.

Growth Mindset
The belief that one's basic abilities can be developed through dedication and hard work; brains and talent are merely the starting point. These types love learning and tend to embrace failure.

Inductive Learning
A reverse method of learning in which students or team-members are presented with a challenge and then learn what they need to know to address the challenge.

MVP
An acronym for Minimum Viable Product, a product with just enough features to satisfy early customers, and to provide feedback for future product development.

Operational Capability
The ability to align critical processes, resources, and technologies in accordance with the overall guiding vision and value proposition of a company or firm.

Project Scoping
Determining and documenting a list of specific project goals, deliverables, tasks, costs, and deadlines to better understand the project as a whole.

Psychology of Innovation
The mindset and behaviors that are conducive to creative and divergent thinking in the context of a business or organizational venture.

Value Proposition
An innovation, service, or feature intended to make a company or product attractive to its customers. Measured by the value a company will deliver to its customers.

Author Biography

Ikhlaq Sidhu, Chief Scientist, and Faculty Director of UC Berkeley's Sutardja Center for Entrepreneurship & Technology. Dr. Sidhu developed the Data X-lab and Data-X Course at Berkeley. He holds 75 patents for internet communication technologies. He co-created the Berkeley Method of Entrepreneurship and developed the Entrepreneurship & Technology area at the #1 public university in the world.

In 2018, Dr. Sidhu received the IEEE Major Education Innovation Award. He received Berkeley IEOR Emerging Area Professor Award in 2009 at UC Berkeley. In 1999, he received 3Com Corporation's Inventor of the Year Award. He serves on many advisory boards. Dr. Sidhu received his bachelor's degree in electrical and computer engineering from the University of Illinois at Urbana-Champaign, and his masters' degree and doctorate in electrical engineering from Northwestern University.

Made in the
USA
Middletown, DE